SpringerBriefs in Electrical and Computer Engineering

SpringerBriefs in Speech Technology

Series Editor:
Amy Neustein

For further volumes:
http://www.springer.com/series/10059

Editor's Note

The authors of this series have been hand selected. They comprise some of the most outstanding scientists—drawn from academia and private industry—whose research is marked by its novelty, applicability, and practicality in providing broad-based speech solutions. The Springer Briefs in Speech Technology series provides the latest findings in speech technology gleaned from comprehensive literature reviews and *empirical investigations* that are performed in both laboratory and *real life* settings. Some of the topics covered in this series include the presentation of real life commercial deployment of spoken dialog systems, contemporary methods of speech parameterization, developments in information security for automated speech, forensic speaker recognition, use of sophisticated speech analytics in call centers, and an exploration of new methods of soft computing for improving human–computer interaction. Those in academia, the private sector, the self service industry, law enforcement, and government intelligence are among the principal audience for this series, which is designed to serve as an important and essential reference guide for speech developers, system designers, speech engineers, linguists, and others. In particular, a major audience of readers will consist of researchers and technical experts in the automated call center industry where speech processing is a key component to the functioning of customer care contact centers.

Amy Neustein, Ph.D., serves as editor in chief of the *International Journal of Speech Technology* (Springer). She edited the recently published book *Advances in Speech Recognition: Mobile Environments, Call Centers and Clinics* (Springer 2010), and serves as quest columnist on speech processing for Womensenews. Dr. Neustein is the founder and CEO of Linguistic Technology Systems, a NJ-based think tank for intelligent design of advanced natural language-based emotion detection software to improve human response in monitoring recorded conversations of terror suspects and helpline calls.

Dr. Neustein's work appears in the peer review literature and in industry and mass media publications. Her academic books, which cover a range of political, social, and legal topics, have been cited in the Chronicles of Higher Education and have won her a pro Humanitate Literary Award. She serves on the visiting faculty of the National Judicial College and as a plenary speaker at conferences in artificial intelligence and computing. Dr. Neustein is a member of MIR (machine intelligence research) Labs, which does advanced work in computer technology to assist underdeveloped countries in improving their ability to cope with famine, disease/illness, and political and social affliction. She is a founding member of the New York City Speech Processing Consortium, a newly formed group of NY-based companies, publishing houses, and researchers dedicated to advancing speech technology research and development.

Dia AbuZeina • Moustafa Elshafei

Cross-Word Modeling for Arabic Speech Recognition

Dia AbuZeina
King Fahd University
of Petroleum and Minerals
Dhahran, Saudi Arabia
abuzeina@hotmail.com

Moustafa Elshafei
King Fahd University
of Petroleum and Minerals
Dhahran, Saudi Arabia
elshafei@kfupm.edu.sa

ISSN 2191-8112 e-ISSN 2191-8120
ISBN 978-1-4614-1212-0 e-ISBN 978-1-4614-1213-7
DOI 10.1007/978-1-4614-1213-7
Springer New York Dordrecht Heidelberg London

Library of Congress Control Number: 2011943295

Printed on acid-free paper

Springer is part of Springer Science+Business Media (www.springer.com)

Preface

The fast pace of the advancement in information and communications technology is reshaping our society and vastly increasing our capabilities for faster learning, higher achievements, and better and wider communication, in addition to more effective and productive collaboration among speech scientists and engineers.

One of the important frontiers of communication technology is the user interface, namely, how can the man–machine interface be designed both in a more natural environment and a more immersive environment, which captures the essential attributes of a human-like exchange between human and machine. To address this important issue, researchers from various areas have been hard at work to equip machines with vital human-like capabilities, such as speech communication and vision. It is fair to say that despite many staggering technological successes achieved in these areas, the machine capabilities developed so far remain rather primitive compared to that of their human counterparts. This has propelled speech system designers to continue their relentless effort to achieve this far-reaching goal.

One such general area where research is continuing persistently is the speech processing area. Speech is the natural form of communication between humans. Its production is a highly nonlinear process that is strongly influenced by the high variability of factors such as age, gender, rate of speech, different dialects and regional accents, emotional state, and more. Speech perception is a hard task in that, in addition to the above-cited production-related difficulties, it has to contend with other equally variable and adverse factors such as background noise, interference from other speakers, room acoustics, recording equipment, and channel characteristics in the case of telephone conversation. Automatic speech recognition (ASR) is a key technology for a variety of applications, such as automatic translation, hands-free operation and control (as in cars and airplanes), automatic query answering, telephone communication with information systems, automatic dictation (speech-to-text transcription), and government information systems. In fact, speech communication with computers and household appliances is envisioned to be the dominant human–machine interface in the near future. However, despite many impressive achievements in the area of speech recognition, reaching well-functioning human performance levels still remains a possibly unattainable goal.

During the last few decades, much research was carried out in the ASR area resulting in numerous practical and commercial successes with impressive high recognition performances, but only if the environment and the speaking manner are constrained such as with the use of isolated keywords.

No doubt, conversational or continuous speech recognition introduces many challenges to ASRs. One of these challenges is the pronunciation variation problem which known to reduce recognition accuracy. Pronunciation variation appears in the form of insertions, deletions, or substitutions of phoneme(s) relative to the canonical transcription of the words in the pronunciation dictionary. Within-word variations and cross-word variations (words' junctures merging) are well-known variation problems in continuous speech. Accordingly, handling this phenomenon is a major requirement to have robust ASRs.

Within-words variation can be accounted for in ASR by using multiple pronunciation variants in the pronunciation dictionaries. However, cross-word variations alter the phonetic spelling of words beyond their listed forms in the pronunciation dictionary, leading to a number of out-of-vocabulary (OOV) word forms. This book presents a knowledge-based approach to model cross-word pronunciation variation at both pronunciation dictionary and language model levels. The proposed approach is based on modeling cross-word pronunciation variations by expanding the pronunciation dictionary and transcription corpus using modern standard Arabic (MSA) phonological rules.

The proposed method was tested using a baseline system that contains a pronunciation dictionary of 14,234 words from a 5.4-h pronunciation corpus of Arabic broadcast news. The expanded dictionary contains 16,273 words. Also, the corpus transcription is expanded according to the applied Arabic phonological rules. Using Carnegie Mellon University (CMU) Sphinx-III speech recognition engine, the enhanced system achieved a word error rate (WER) of 9.91% on a test set of fully diacritized transcription of about 1.1 h of MSA broadcast news. The WER is significantly reduced by 2.30% compared to the baseline system.

This book presents many examples in Arabic; a full appendix is provided for the Arabic terminologies used in this book.

Dhahran, Saudi Arabia Dia AbuZeina
Dhahran, Saudi Arabia Moustafa Elshafei

Acknowledgments

The authors would like to thank King Fahd University of Petroleum and Minerals for providing the facilities to write this book.

Contents

Chapter 1
An Overview of Speech Recognition Systems

This chapter presents an introduction to automatic speech recognition systems. It includes the mathematical formulation of speech recognizers. The main components of speech recognition systems are introduced: Front-end signal processing, acoustic models, decoding, training, language model, and pronunciation dictionary. Additionally, a brief literature review of speech recognition systems is also provided. Viterbi and Baum–Welch algorithms are also discussed as the fundamental techniques for decoding and training phases, respectively.

1.1 Introduction

A speech recognizer is a program that transcribes speech into texts for many purposes; facilitating human computer interface is the major advantage. A wider reach of the information technology in the society can be achieved if users can verbally communicate with computer. In fact, being able to speak fluently with computer eliminates handwriting and spelling problems as well as having words spelled incorrectly, and increases the productivity of people. Nowadays, big companies utilize this technology to automate their processes. With huge number of customers, companies offer their services more smoothly as a user can verbally inquire, order, and pay. In addition to the commercial applications, speech recognition is also employed in eLearning, training, and education of students with learning disabilities Khasawneh et al. (2004) listed some speech recognition applications, which include banking by telephone, automatic teller machines, compact size computers, browsing computer networks and databases by voice, and operating machinery from a distance in dangerous working sites. However, there are also some drawbacks. Speech recognition systems require high computational machines with large memory. In addition, high rate of misrecognitions and errors is still a major problem in speech recognition systems, and hinders its widespread adaptation in the IT applications.

D. AbuZeina and M. Elshafei, *Cross-Word Modeling for Arabic Speech Recognition*, SpringerBriefs in Electrical and Computer Engineering,
DOI 10.1007/978-1-4614-1213-7_1, © Dia AbuZeina 2012

Benzeghiba et al. (2007) presented a comprehensive study on pronunciation variations as major sources of errors in automatic speech recognition (ASR). They showed some of the speech variability sources: foreign and regional accents, speaker physiology, speaking style and spontaneous speech, rate of speech, children speech, emotional state, and more.

A typical large vocabulary speech recognizer would first convert speech waveform into a sequence of feature vectors then identify the phones (basic pronunciation units) corresponding to the feature sequence, and finally transcribe the recognized phone strings into a sequence of words. Of the numerous approaches to solve the ASR problem, two major ones stand out (Huang et al. 2001; Rabiner and Juang 1993): the knowledge-based and data-based statistical approach. In the former approach, the aim is to derive explicit rules from the human knowledge of speech acquired by human experts. Examples of such rules are acoustic-phonetic rules, lexicographic rules (i.e., rules describing words in the lexicon), and syntactic rules exploiting the syntax of the language. This technique is firmly rooted in the areas of artificial intelligence and expert systems. Although this approach has led to a limited number of practical and commercial successes, its use requires large databases, intensive computational resources, and a great deal of expertise, and entails eliciting a number of rules that tend to increase with the required recognition rate. As such, this approach did not seem to have caught on. In contrast, the data-driven statistical approach has virtually dominated ASR research over the last few decades. The statistical approach is itself dominated by the powerful statistical technique called Hidden Markov Model (HMM). Based on the pioneering work of Jelinek (1998) and Baker (1975), the HMM-based ASR technique has led to numerous successful applications requiring large vocabulary speaker-independent continuous speech recognition (Morgan and Bourlard 1995; Young 1996).

The HMM-based technique essentially consists of recognizing speech by estimating the likelihood of each phoneme at contiguous, small frames of the speech signal (Rabiner 1989; Rabiner and Juang 1993). Words in the target vocabulary are modeled into a sequence of phonemes and then a search procedure is used to find, among the words in the vocabulary list, the phoneme sequence that best matches the sequence of phonemes of the spoken word. Each phoneme is modeled as a sequence of HMM states. In standard HMM-based systems, the likelihoods (also known as the emission probabilities) of a certain frame observation being produced by a state are estimated using traditional Gaussian mixture models. The use of HMM with Gaussian mixtures has several notable advantages such as a rich mathematical framework, efficient learning and decoding algorithms, and an easy integration of multiple knowledge sources.

Two notable successes in the academic community in developing high performance large vocabulary, speaker-independent speech recognition systems are the HMM tools, known as the Hidden Markov Model Toolkit (HTK), developed at Cambridge University (Young et al. 2004; HTK 2011), and the Sphinx system developed at Carnegie Mellon University (CMU) (Lee 1988; CMU Sphinx 2011). HTK is a general purpose toolkit for building HMMs and has been used in many applications. On the contrary, CMU Sphinx system was built specifically for speech recognition applications. In this study, we use Sphinx-based ASR system for testing and evaluation.

The Sphinx Group at CMU has been supported for many years by funding from the Defense Advanced Research Projects Agency (DARPA) and industries to assess and develop speech recognition techniques. In 2000, the Sphinx Group released Sphinx-II, a real-time, large vocabulary, speaker-independent speech recognition system as free software under the Apache-style license. The source code is freely available for educational institutions. The extensive source code resources represent an excellent research infrastructure and a powerful test bed for researchers to pursue further state-of-the-art research in the area of speech recognition techniques. CMU Sphinx toolkit has a number of packages for different tasks and applications (Open Source Toolkit for Speech Recognition 2011) Some examples are as follows:

- Pocketsphinx—recognizer library written in C
- Sphinxbase—support library required by Pocketsphinx
- Sphinx4—adjustable, modifiable recognizer written in Java
- CMUclmtk—language model tools
- SphinxTrain—acoustic model training tools

And the latest available releases are as follows:

- sphinxbase-0.7
- pocketsphinx-0.7
- sphinx4-1.0beta6
- sphinxtrain-1.0.7
- cmuclmtk-0.7

The statistical approach using HMM has been the dominant technique for speech recognition systems for the last two decades. HMM-based speech recognition systems started around 1975 when James Baker applied statistical method to speech recognition (Baker 1975). In fact, we do not need to mention the research work since that date; instead we present a brief literature to highlight the recent works and advancements in different components of HMM-based ASRs. Rabiner and Juang (2004) outlined the major components of a HMM-based modern speech recognition and spoken language understanding system. Bilmes (2006) presented a list of possible HMM properties in terms of random variables and conditional independence assumptions. His effort was to introduce a better understanding of what HMMs can do. Baker et al. (2007) presented a report to survey historically significant events in speech recognition and understanding which have enabled this technology to become progressively more capable and cost effective in a growing number of everyday applications. Deng and Huang (2004) demonstrated a number of fundamental and practical limitations in speech recognition technology which hinder ubiquitous adoption of this widely used technology. Hong-Kwang Jeff and Yuqing (2006) presented a framework for speech recognition using maximum entropy direct modeling, where the probability of a state or word sequence given an observation sequence is computed directly from the model. In contrast to HMMs, features can be asynchronous and overlapping. Ye-Yi et al. (2008)

categorized spoken dialog technology into form filling, call routing, and voice search, and reviewed the voice search technology. Dong et al. (2008) described a new nonlinear noise reduction algorithm motivated by the minimum mean-square-error (MMSE) criterion in the mel-frequency cepstral coefficients (MFCCs) domain for environment-robustness ASR. Zweig and Nguyen (2009) proposed a segmental conditional random fields (CRF) approach to large vocabulary continuous speech recognition systems. They achieved improvement of 2% compared to the HMM-based baseline. ASRs enhancements also include the linguistic parts. Luo (2011) proposed an improved speech recognition algorithm based on a hybrid support vector machine (SVM) and HMM architecture. The experimental results showed that the recognition rate had increased greatly. To overcome the flaws of the HMM paradigm, Xi et al. (2005) designed a hybrid HMM/artificial neural networks (ANN) model. In this hybrid model, the nonparametric probabilistic model (a BP neural network) is used to substitute the Gauss blender to calculate the observed probability that is necessary for computing the states of the HMM. Xiao and Qin (2010) demonstrated that feature coefficients based on MFCC are not fully reflecting speech information as a result of speech signal movement and overlap of frames, especially noisy effect. They presented a new method for noise robust speech recognition based on a hybrid model of HMM and Wavelet Neural Network (WNN). Their experimental results show a better noise robustness model. Sloin and Burshtein (2008) presented a discriminative training algorithm that uses SVMs, to improve the classification of discrete and continuous output probability HMMs. The presented algorithm uses a set of maximum-likelihood (ML)-trained HMMs as a baseline system, and an SVM training scheme to rescore the results of the baseline HMMs. Xian (2009) presented the use of a hybrid HMM and ANNs for ASR. The proposed hybrid system for ASR was to take advantage from the properties of both HMM and ANN, improving flexibility and recognition performance. Middag et al. (2009) presented a novel methodology that utilizes phonological features to assess the pathological state of the speaker using ASR. Cao et al. (2005) proposed an approach to extend the existing language modeling approach by relaxing the independence assumption. They proposed two types of relationship extracted from WordNet and co-occurrence relationships respectively. Salgado-Garza et al. (2004) used tag trigrams as a post-processing stage. They demonstrated small but statically significant improvements in the recognition accuracy of the Sphinx 3 decoder. Beutler (2007) demonstrated a method to bridge the gap between statistical language models and elaborate linguistic grammars. He introduced precise linguistic knowledge into a medium vocabulary continuous speech recognizer. His results showed a statistically significant improvement of recognition accuracy on a medium vocabulary continuous speech recognition dictation task. Schwenk (2007) described the use of a neural network language model for large vocabulary continuous speech recognition. The underlying idea of his approach was to attack the data sparseness problem by performing the language model probability estimation in a continuous space. Yuecheng et al. (2008) suggested using a gating network to modulate the effects of the context to improve the performance of a neural network language model. It was found that it is a very effective way.

1.2 Speech Recognition Architectures

Modern large vocabulary, speaker-independent, continuous speech recognition systems have three knowledge sources: acoustic models, language model, and pronunciation dictionary (also called lexicon). A lexicon provides pronunciation information for each word in the vocabulary in phonemic units, which are modeled in detail by the acoustic models. The language model provides the a priori probabilities of word sequences. Of course, there are other parts such as front-end and decoder which are often included in the architecture. Figure 1.1 shows Sphinx engine architecture.

The figure illustrates the subsystems available in Sphinx tools and the relationship between them. The following is a brief description of the main subfunctions of Sphinx engine:

- The front-end: This subsystem provides the initial step in converting sound input into a form of data usable by the rest of the system (called features). Sphinx provides different front-ends, and allows the users of the framework to select whichever front-end they desire, or even develop their own front-end. The front-end used in our analysis is the MFCC.
- The Linguist: This subsystem contains the details that describe the recognized language itself (or the necessary subset of this language). This subsystem is where most of the adjustments will be made in order to support the Arabic language. It consists of three main modules:
  - *The Acoustic Model*. This module provides the HMMs that can be used to recognize speech. Sphinx provides various acoustic models.

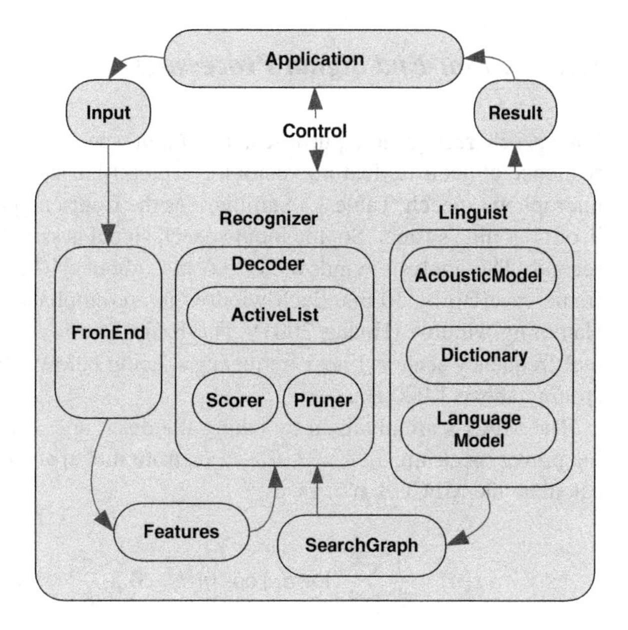

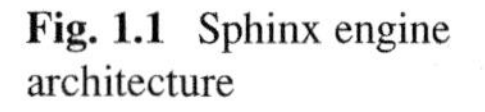
Fig. 1.1 Sphinx engine architecture

- *The Language Model*. This module provides the grammar that is used by the system (usually the grammar of a natural language or a subset of it).
- *The Dictionary*. This module serves as an intermediary between the Acoustic Model and the Language Model. It contains the words available in the language and the pronunciation of each in terms of the phonemes available in the acoustic model.

- The decoder: This subsystem does the actual recognition job. When speech is entered into the system, the front-end converts it into features as described earlier. The decoder takes these features, in addition to the search graph provided by the Linguist, and tries to recognize the speech in the features. In testing steps the decoder requires acoustic models, a pronunciation dictionary and a language model.

The speech recognition problem is to transcribe the most likely spoken words given the acoustic observations. If $O = [o_1, o_2, \ldots, o_n]$ is the acoustic observation and $W = [w_1, w_2, \ldots, w_n]$ is a word sequence, then

$$\hat{W} = \underbrace{\arg\max}_{\text{for all words}} P(W)P(O|W),$$

where $\hat{W}$ is the most probable transcription of the spoken words which is also called maximum posteriori probability. $P(W)$ is the prior probability computed in the language model, and $P(O|W)$ is the probability of observation likelihood computed using the acoustic model. The following sections briefly present the main parts of typical speech recognition architecture.

1.2.1 Front-End Signal Processing

For speech recognition purposes, the input speech signal is transferred into a sequence of acoustic feature vectors. A 16,000-Hz sampling rate is often used for microphone speech. Table 1.1 summarizes the front-end parameters which are used in our baseline system. So, the input speech signal is sampled at 16,000 samples per second. The analysis window is 25.6 ms (about 410 samples), and consecutive frames overlap by 10 ms. Each window is pre-emphasized and is multiplied by a Hamming window (Huang 2001). The basic feature vector uses the MFCC. The mel-frequency scale is linear frequency spacing below 1,000 Hz and a logarithmic spacing above 1,000 Hz.

The MFCCs are obtained by taking the discrete cosine transform (DCT) of the log power spectrum, S_k, $k = 1, 2, \ldots, K$, from mel-spaced filter banks. We can then calculate the MFCC's $x(n)$ as

$$x(n) = \frac{1}{L} \sum_{k=1}^{k=K} (\log S_k) \cos\left[n(k-0.5)\frac{\pi}{K}\right], \quad n = 1, 2, \ldots, L. \qquad (1.1)$$

Table 1.1 Front-end signal processing parameters

No.	Parameters	Sphinx
1	Sampling rate	16,000 samples per second
2	Quantization level	16 bits per sample
3	Frame rate	100 frames per second
4	Window length	0.0256 s
5	Filter bank type	Mel-frequency filter bank
6	Number of cepstra	13
7	Number of mel filters	40
8	DFT size	256
9	Lower frequency	133.33 Hz
10	Upper frequency	6,855.5 Hz
11	Pre-emphasize	0.97
12	Dimension of the basic MFCC feature vector	13
13	Dimension of the overall feature vector	39

Sphinx system uses a 13-coefficients basic feature vector, $x_t(k)\, 0 \leq k \leq 12$.

The basic feature vector is usually normalized by subtracting the mean over the sentence utterance. $x(0)$ represents the log mel spectrum energy, and is used to derive other feature parameters. The basic feature vector is highly localized. To account for the temporal properties, two other derived vectors are constructed from the basic MFCCs: a 40- and a 80-ms differenced MFCC (13 parameters), a 13-coefficient second-order differenced MFCC giving a feature vector dimension of 39 (Lee 1988; Lee et al. 1990). Alternative feature vectors of four streams are also commonly used in when vector quantization is used to simplify the calculations of the emission probabilities. In our system, a single feature vector, xt, of 39 parameters per frame is used throughout training and decoding. In the following discussion, we will refer to the feature vector at time t by xt. The feature vector is also called the observation vector.

1.2.2 Acoustic Model

Precise acoustic model is a key factor to improve recognition accuracy as it describes the HMM of the phonemes (the basic units of sounds). Sphinx uses 39 English phones (The CMU Pronunciation Dictionary 2011). The acoustic model uses a 3- to 5-state Markov chain to represent the speech phoneme (Lee 1988). Figure 1.2 is an acoustic model to represent a phoneme which is also known as Bakis model. The illustrated Bakis model has a fixed topology consisting of an input state, three emitting states, two branches, and an output state.

An HMM model, $\boldsymbol{\lambda}$, is completely specified by the following parameters (Rabiner 1989):

- The number of states N.
- The state transition probabilities, A, $a_{ij} = P(s_{t+1} = j | s_t = i)$, where s_t is the state at time t.

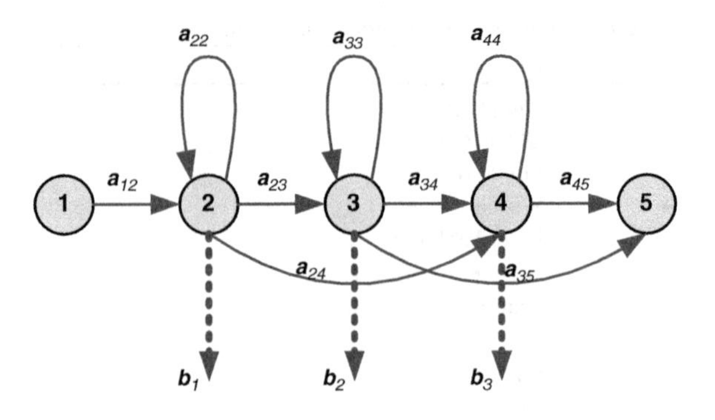

Fig. 1.2 Bakis HMM

- The observation symbol probability, B, $b_j(x_t) = P(x_t|s_t = j)$, where x_t is the observation at time t.
- The initial state probabilities, Π, $\pi_{\mathrm{i}} = P(s_1 = i)$.

The transition matrices of these models are estimated by considering a single model for each phoneme, without considering its context. On the contrary, the calculation of the observation symbol probability (also called the emission probability, or the state output probability) requires further elaboration. The observation probability is highly dependent on the context of a phoneme. Each phone is influenced to different degrees by its neighboring phones. For better acoustic modeling, Sphinx uses triphones. Triphones are context-dependent models of phonemes, each triphone represents a phone surrounded by specific left and right phones (Hwang 1993). For example the phoneme /T/ when /EY/ appears on its left and /L/ appears on its right is the triphone /T(EY,L)/. For example, in the DARPA Resource Management (RM) speech corpus (Price et al. 1988), a 1,000 word corpus vocabulary, there were 7,549 English triphones. Each phoneme HMM uses different distributions to model the observation probability of its triphones. Sphinx uses two different techniques for parametrizing the probability distributions of the state emission probabilities: continuous HMM (CHMM) and semicontinuous HMM (SCHMM) (Lamere et al. 2003; Huang 1992; Gauvain and Lee 1994). The semicontinuous requires substantially a smaller number of parameters and is faster in decoding, but is only good for limited vocabulary. The continuous one uses more parameters and is slower in decoding, but proves to be successful for large vocabulary applications.

In CHMM, for example, the Gaussian mixture density is used, and the probability of generating the observation $\boldsymbol{x}_t$ given the transition state j, $P(x_t|j)$ becomes

$$b_j(x_t) = p(x_t|q_t = j) = \sum_{k=1}^{M} w_{j,k} N_{j,k}(x_t), \tag{1.2}$$

where $N_{j,k}$ is the kth Gaussian distribution, $w_{j,k}$ are the mixture weights, and $\sum_k w_{j,k} = 1$. CHMM is the most popular method today for large vocabulary speech recognition systems. However, its main drawback is the extremely large number of parameters needed to describe the Gaussian distributions. For example for the RM applications, if we use $M = 16$, and three states for each HMM, the number of distributions would be $16 \times 3 \times 7{,}549$. Moreover, each distribution requires two 39-element vectors for its mean and variance parameters.

Reducing the number of parameters to describe all the acoustic models of all triphones can be achieved by using the concept of shared distributions (Hwang and Huang 1993). In this technique all the states of all triphones of a given phoneme share a common pool of probability distributions. These shared distributions are called senones. For example, in one of the Sphinx-III applications there were 4,500 senons. The phoneme /AE/ had 138 triphones, but the emission probabilities of the states (five states for each triphone) share a pool of only 127 distributions. The values of $\boldsymbol{b}_j(\boldsymbol{x}_t)$ are pre-computed as discrete probabilities and stored in lookup tables.

1.2.3 Decoding Using Viterbi Algorithm

A basic step in recognition is to calculate the probability of observing a sequence of speech features $X = \{x_1, x_2, \ldots, x_T\}$, given a phoneme HMM, $\boldsymbol{\lambda}$, $P(X|\lambda)$. We then need to enumerate every possible state sequence of length T.

Consider the sequence $S = [s_1, s_2, \ldots s_T]$, the probability of observing such sequence of feature vectors given the model is obtained by summing up all possible state sequences of length T.

$$P(X|\lambda) = \sum_{\text{all} S} P(X|S, \lambda) P(S|\lambda),$$

$$P(X|\lambda) = \sum_{\text{all } S} \pi_{s1} b_{s1}(x_1) \prod_{t=2}^{T} a_{st-1,st} b_{st}(x_t). \tag{1.3}$$

Equation (1.3) can be efficiently calculated using an iterative procedure called forward–backward procedure. Isolated word recognition or recognition of limited number of sentences can be performed by selecting the model of the sentence which gives the highest probability of observations. In large vocabulary system, where there could be large possibilities of phoneme sequences, a recognition procedure is needed for matching the observed sound wave with the nearest sequence of phones.

Viterbi algorithm is used to find the highest scoring state sequence, $q = (s_1, s_2, \ldots, s_T)$ for a given observation sequence $X = (x_1, x_2, \ldots, x_t, \ldots, x_T)$, i.e., find $S_{\text{best}} = \mathbf{arg}\{\mathbf{\max_S} P(S|X)\}$,

which is equal to

$$\arg\left\{\max_S \prod_{i=1,\ldots,K} P(x_i|s_i, s_{i-1}) p(s_i|s_{i-1})\right\}. \tag{1.4}$$

Let us define $\phi(t,i)$ to be the probability of the most likely partial state sequence or path until time t, and ending at the ith state, the algorithm proceeds in the following steps (Rabiner and Juang 1993; Rabiner 1989; Forney 1973):

Step 1: Initialization

$$\phi(1,j) = a_{1,j} b_j(x_1). \tag{1.5}$$

Step 2: Induction

$$\phi(t,j) = \max_i \{\phi(t-1,i)a_{i,j}\} b_j(x_t), \quad i = 1,2,\ldots,N \text{ and } t = 2,3,\ldots,T, \tag{1.6}$$

$$U(t,i) = \arg\left\{ \max_j \{\phi(t-1,j)a_{i,j}\} b_j(x_t) \right\},$$
$$j = 1,2,\ldots,N \text{ and } t = 2,3,\ldots,M. \tag{1.7}$$

Step 3: Best path: the maximum likelihood of the best path is then given by

$$P(X|\text{Model}) = \varphi(N,T) = \left\{ \max_j \{\phi(N,j)\} \quad j = 1,2,\ldots,n_v(M) \right\},$$

$$U(M,i_{\text{best}}) = \arg\left\{ \max_j \{\phi(M,j)\} \quad j = 1,2,\ldots,n_v(M) \right\}. \tag{1.8}$$

Step 4: Backtracking

$$\begin{aligned} i_M &= i_{\text{best}} \\ i_{t-1} &= U(t,i_t) \quad \text{for } t = M, M-1,\ldots,2. \\ S &= s_{i1} s_{i2} \ldots s_{iM} \end{aligned} \tag{1.9}$$

1.2.4 Training Using Baum–Welch Algorithm

Training speech recognition system consists of building two models; the language model and the acoustic model. In natural language speech recognition system, the language model is a statistically based model using unigram, bigrams, and trigrams of the language for the subject text to be recognized. On the contrary, the acoustic model builds the HMMs for all the triphones and the probability distribution of the observations for each state in each HMM.

Sphinx training tools have a set of executables and Perl scripts that cooperate to create acoustic models for Sphinx speech applications. The models can be built and configured directly using the provided scripts, or by manually running the executables.

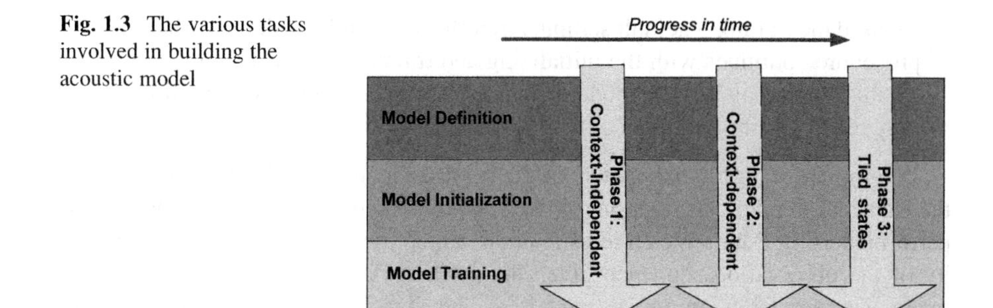

Fig. 1.3 The various tasks involved in building the acoustic model

The training process for the acoustic model consists of three phases, as shown in Fig. 1.3. Each phase consists of three stages (model definition, model initialization, and model training) and makes use of the output of its previous phase. The phases are as follows:

- Context-independent phase (CI): The context-independent phase creates a single HMM for each phoneme in the phone list. The number of states in an HMM can be specified by the developer; in the model definition stage, a serial number is assigned for each state in the whole acoustic model. Additionally, the main topology for the HMMs is created. The topology of an HMM specifies the possible state transitions in the acoustic model, the default is to allow each state to loop back and move to the next state; however, it is possible to allow the states to skip to the second next state directly. In the model initialization, some model parameters are initialized to some calculated values. The model training stage consists of a number of executions of the Baum–Welch algorithm (5–8 times) followed by a normalization process.
- Untied context-dependent phase (CD): In this phase, triphones are added to the HMM set. In the model definition stage, all the triphones appearing in the training set will be created, and then the triphones below a certain frequency are excluded for the purposes of this phase. Specifying a reasonable threshold for frequency is important for the performance of the model.

  After defining the needed triphones, states are given serial numbers as well (continuing the same count). The initialization stage copies the parameters from the CI phase. Similar to the previous phase, the model training stage consists of number of executions of the Baum–Welch algorithm (6–10 times) followed by a normalization process.
- Tied context-dependant phase: This phase aims to improve the performance of the model generated by the previous phase by tying some states of the HMMs. These tied states are called senones. The process of creating these senones involves building some decision trees based on some "linguistic questions" provided by the developer. For instance, these questions could be about the classification of phonemes according to some acoustic property. If the user did not supply these questions, SphinxTrain could guess these questions by analyzing the voice transcriptions provided in the training data. At this stage,

we used the Sphinx 3 default setting. After the new model is defined, the training procedure continues with the initializing and training stages. The training stage for this phase may include modeling with a mixture of normal distributions. This may require more iterations of the Baum–Welch algorithm.

Determination of the parameters of the acoustic model is referred to as training the acoustic model. Estimation of the parameters of the acoustic models is performed using Baum-Welch re-estimation, which tries to maximize the probability of the observation sequence given the model. The algorithm proceeds iteratively, starting from an initial model λ. The steps in this algorithm may be summarized as follows:

Step 1: Calculate the forward and backward probabilities for all states j and times t.

Step 2: Update the parameters of the new model as follows:

$$\bar{\pi}_j = \text{expected frequency of the state j at time t} = 1, \tag{1.10}$$

$$\bar{a}_{ij} = \frac{\text{Expected number of transition from state } i \text{ to state } j}{\text{Expected number of transitions from state } i}, \tag{1.11}$$

$$\bar{b}_j(k) = \frac{\text{expected number of times in state } j \text{ and observation symbol } x_k}{\text{expected number of times in state } j}. \tag{1.12}$$

If for each state the output distribution is a single component Gaussian, the parameters of the distribution can be found

$$\bar{\mu}_j = \frac{\sum_{t=1}^{T} L_j(t) x_t}{\sum_{t=1}^{T} L_j(t)}$$

is the mean value of the observation vectors emitted at state j.

$$\bar{\Sigma}_j = \frac{\sum_{t=1}^{T} L_j(t)(x_t - \bar{\mu}_j)(x_t - \bar{\mu}_j)'}{\sum_{t=1}^{T} L_j(t)}$$

is the covariance matrix of the observation vectors emitted at state j, where $L_j(t)$ is probability of being in state j at the time t, given the observation sequence and the model.

Step 3: If the value of $P(X|\lambda)$ for this iteration is not higher than the value at the previous iteration, then stop. Otherwise, repeat the above steps using the new re-estimated parameter values.

1.2.5 Language Model

Speech recognition systems treat the recognition process as one of maximum a posteriori estimation, where the most likely sequence of words is estimated, given the sequence of feature vectors for the speech signal. Mathematically, this can be represented as (Huang 2001)

$$\text{Word1Word2Word3}\ldots = \arg\max_{\text{Wd1Wd2}\ldots}\{P(\text{feature vectors}|\text{Wd1Wd2}\ldots)P(\text{Wd1Wd2}\ldots)\}, \tag{1.13}$$

where Word1Word2... is the recognized sequence of words and Wd1Wd2... is any sequence of words. The argument on the right-hand side of (1.1) has two components: the probability of the feature vectors, given a sequence of words *P*(feature vectors|Wd1Wd2...), and the probability of the sequence of words itself, *P*(Wd1Wd2...). The first component is provided by the acoustic model. The second component, also called the language component, is provided by a language model. The most commonly used language models are N-gram language models. These models assume that the probability of any word in a sequence of words depends only on the previous N words in the sequence. Thus, a bigram language model would compute *P*(Wd1Wd2...) as

$$P(\text{Wd1Wd2Wd3Wd4}\ldots) = P(\text{Wd1})P(\text{Wd2}|\text{Wd1})P(\text{Wd3}|\text{Wd2})P(\text{Wd4}|\text{Wd3})\ldots \tag{1.14}$$

Similarly, a trigram model would compute it as

$$\text{P(Wd1Wd2Wd3}\ldots) = \text{P(Wd1)P(Wd2}|\text{Wd1)P(Wd3}|\text{Wd2, Wd1)P(Wd4}|\text{Wd3, Wd2)}\ldots \tag{1.15}$$

The CMU statistical language tool is described in Clarkson and Rosenfeld (1997). The CMU statistical language tool kit is used to generate our Arabic statistical language model. The steps for creating and testing the language model, shown in Fig. 1.4, are as follows:

- Compute the word unigram counts.
- Convert the word unigram counts into a vocabulary list.
- Generate bigram and trigram tables based on this vocabulary.

The tool generates the language model in two formats: a binary format to be used by the Sphinx decoder, and a portable text file in the standard ARPA format.

The language modeling tools also come with a tool for evaluating the language model. The evaluation measures the goodness of the language model in terms of the perplexity of metric. The ARPA format is the language model type that is used when measuring the perplexity.

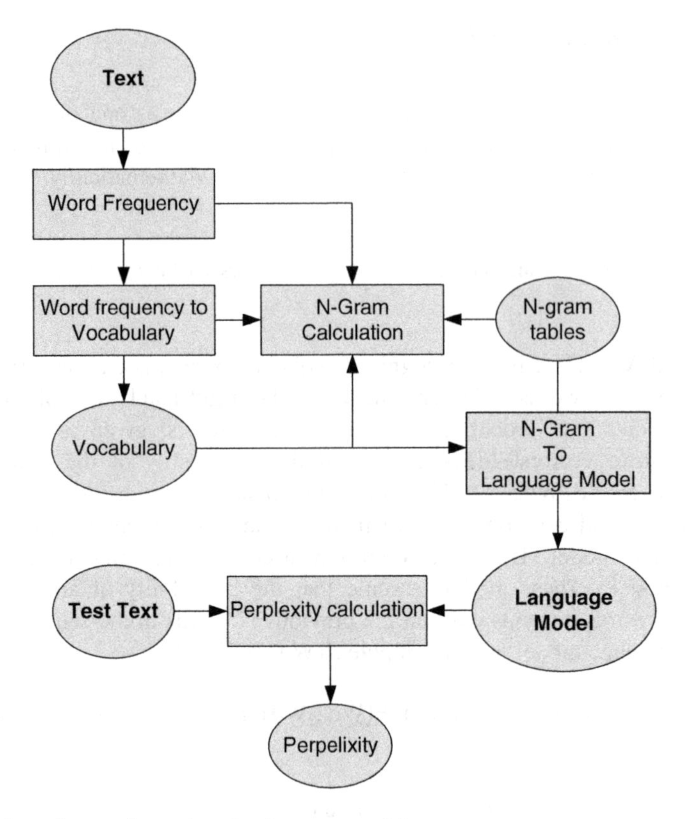

Fig. 1.4 Steps for creating and testing language model

1.2.6 Pronunciation Dictionary

Both training and recognition stages require a pronunciation dictionary which is a mapping table that maps words into sequences of phonemes. A pronunciation dictionary is basically designed to be used with a particular set of words. It provides the pronunciation of the vocabulary for the transcription corpus using the defined phoneme set. Like acoustic model and language model, the performances of the speech recognition systems depend critically on the definitions of the phonemes and the accuracy of the dictionary. In decoding stage, the dictionary serves as intermediary between the acoustic model and the language model. There are two types of dictionary; closed vocabulary and open vocabulary. In closed vocabulary, all corpus transcription words are listed in the dictionary. In contrast, it is possible to have non corpus transcription words in the open vocabulary dictionary. Closed and open vocabularies have a direct effect on the OOV. A phoneme set is manually designed by human expertise. However, when human expertise in not available, it possible to be selected using a data-driven approach as demonstrated by Singh et al. (2002). In addition to providing the words phonetic transcriptions of words of the target vocabulary, the dictionary is the place where a word's multiple pronunciations are added.

References

Baker JK (1975) Stochastic modeling for automatic speech understanding. In: Reddy R (ed) Speech recognition. Academic, New York, pp 521–542

Baker J, Deng L, Glass J, Khudanpur S, Lee C, Morgan N (2007) Historical development and future directions in speech recognition and understanding, MINDS report. http://www-nlpir.nist.gov/MINDS/FINAL/speech.web.pdf

Benzeghiba M, De Mori R et al (2007) Automatic speech recognition and speech variability: a review. Speech Commun 49(10–11):763–786

Beutler R (2007) Improving speech recognition through linguistic knowledge. Doctoral dissertation, ETH Zurich

Bilmes J (2006) What HMMs can do. IEICE Trans Inf Syst E89-D(3):869–891

Cao G, Nie J-Y, Bai J (2005) Integrating word relationships into language models. In: Proceedings of the ACM 28th annual international conference on research and development in information retrieval (SIGIR'05), Salvador, Brazil

Clarkson P, Rosenfeld R (1997) Statistical language modeling using the CMU-Cambridge toolkit. In: Proceedings of the 5th European conference on speech communication and technology, Rhodes, Greece

CMU Sphinx Downloads (2011) http://cmusphinx.sourceforge.net/wiki/download. Accessed 1 Sep 2011

Deng L, Huang X (2004) Challenges in adopting speech recognition. Commun ACM 47(1):69–75

Dong Y, Li D et al (2008) Robust speech recognition using a cepstral minimum-mean-square-error-motivated noise suppressor. IEEE Trans Audio Speech Lang Process 16(5):1061–1070

Forney GD (1973) The Viterbi algorithm. Proc IEEE 61(3):268–278

Gauvain J-L, Lee C-H (1994) Maximum a posteriori estimation for multivariate Gaussian mixture observations of Markov chains. IEEE Trans Speech Audio Process 2(2):291–298

Hong-Kwang Jeff K, Yuqing G (2006) Maximum entropy direct models for speech recognition. IEEE Trans Audio Speech Lang Process 14(3):873–881

HTK (2011) http://htk.eng.cam.ac.uk/. Accessed 1 Sep 2011

Huang XD (1992) Phoneme classification using semicontinuous hidden Markov models. IEEE Trans Signal Process 40(5):1062–1067

Huang X, Acero A, Hon H (2001) Spoken language processing. Prentice Hall PTR, Upper Saddle River, NJ

Huang X, Acero A, Acero A, Hon H (2001) Spoken language processing: a guide to theory, algorithm, and system development. Prentice Hall, New York

Hwang M-H (1993) Subphonetic acoustic modeling for speaker-independent continuous speech recognition, Ph.D. thesis, School of Computer Science, Carnegie Mellon University

Hwang MY, Huang X (1993) Shared-distribution hidden Markov models for speech recognition. IEEE Trans Speech Audio Process 1(4):414–420

Jelinek F (1998) Statistical methods for speech recognition. MIT, Cambridge, MA

Khasawneh M, Assaleh K et al (2004) The application of polynomial discriminant function classifiers to isolated Arabic speech recognition. In: Proceedings of the IEEE international joint conference on neural networks

Lamere P, Kwok P, Walker W, Gouvea E, Singh R, Raj B, Wolf P (2003) Design of the CMU Sphinx-4 decoder. In: Proceedings of the 8th European conference on speech communication and technology, Geneva, Switzerland, pp 1181–1184

Lee KF (1988) Large vocabulary speaker independent continuous speech recognition: the SPHINX system. Doctoral dissertation, Carnegie Mellon University

Lee KF, Hon HW, Reddy R (1990) An overview of the SPHINX speech recognition system. IEEE Trans Acoust Speech Signal Process 38(1):35–45

Luo X (2011) Chinese speech recognition based on a hybrid SVM and HMM architecture advances in neural networks. In: Liu D, Zhang H, Polycarpou M, Alippi C, He H (eds) ISNN 2011, LNCS 6677. Springer, Berlin, pp 629–635

Middag C, Martens J-P et al (2009) Automated intelligibility assessment of pathological speech using phonological features. EURASIP J Adv Signal Process 2009:1–9

Morgan N, Bourlard H (1995) Continuous speech recognition. IEEE Signal Process Mag 12(3): 25–42

Open Source Toolkit for Speech Recognition (2011) http://cmusphinx.sourceforge.net/wiki/download/. Accessed 1 Sep 2011

Price P, Fisher WM, Bernstein J, Pallett DS (1988) The DARPA 1000-word resource management database for continuous speech recognition. In: Proceedings of the IEEE international conference on acoustics, speech and signal processing, vol 1, pp 651–654

Rabiner LR (1989) A tutorial on hidden Markov models and selected applications in speech recognition. Proc IEEE 77(2):257–286

Rabiner L, Juang B (1993) Fundamentals of speech recognition. Prentice Hall, Upper Saddle River, NJ

Rabiner LR, Juang BH (2004) Statistical methods for the recognition and understanding of speech. In: Encyclopedia of language and linguistics, *Second Edition*, 2005

Salgado-Garza LR, Stern RM, Nolazco FJA. (2004). N-Best list rescoring using syntactic trigrams. In: Monroy R, Arroyo-Figueroa G, Sucar L, Sossa H (eds), MICAI 2004, LNAI 2972, Springer, Berlin, pp 79–88

Schwenk H (2007) Continuous space language models. Comput Speech Lang 21(3):492–518

Singh R, Raj B et al (2002) Automatic generation of subword units for speech recognition systems. IEEE Trans Speech Audio Process 10(2):89–99

Sloin A, Burshtein D (2008) Support vector machine training for improved hidden Markov modeling. IEEE Trans Signal Process 56(1):172–188

The CMU Pronunciation Dictionary (2011) http://www.speech.cs.cmu.edu/cgi-bin/cmudict. Accessed 1 Sep 2011

Xi X, Lin K, Zhou C, Cai J (2005) A new hybrid HMM/ANN model for speech recognition. In: Proceedings of the second IFIP conference on artificial intelligence applications and innovations (AIAI 2005), pp 223–230

Xian T (2009) Hybrid Hidden Markov Model and artificial neural network for automatic speech recognition. Pacific-Asia conference on circuits, communications and systems, 2009. PACCS'09

Xiao Y, Qin A (2010) Noise robust speech recognition based on improved hidden Markov model and wavelet neural network. Comput Eng Appl 46(22): pp 162–164, 235

Ye-Yi W, Dong Y et al (2008) An introduction to voice search. IEEE Signal Process Mag 25 (3):28–38

Young S (1996) A review of large-vocabulary continuous-speech recognition. IEEE Signal Process Mag 13(5):45–57

Young SJ, Evermann G, Gales MJF, Hain T, Kershaw D, Moore GL, Odell JJ, Ollason D, Povey D, Valtchev V, Woodland PC (2004) The HTK Book

Yuecheng Z, Mnih A, Hinton G (2008) Improving a statistical language model by modulating the effects of context words, in: ESANN, 2008

Zweig G, Nguyen P (2009) A segmental CRF approach to large vocabulary continuous speech recognition. IEEE workshop on automatic speech recognition and understanding, 2009. ASRU 2009

Chapter 2
Arabic Speech Recognition Systems

This chapter presents a brief overview of the evolution of Arabic speech recognition systems. It provides a literature survey of Arabic speech recognition systems and discusses some of the challenges of Arabic from the speech recognition point of view.

2.1 Literature and Recent Works

Development of an Arabic speech recognition is a multidiscipline effort, which requires integration of Arabic phonetic (Elshafei 1991; Alghamdi 2000; Algamdi 2003), Arabic speech processing techniques (Elshafei et al. 2002, 2007; Al-Ghamdi et al. 2003), and natural language processing (Elshafei et al. 2006). Development of an Arabic speech recognition system has recently been addressed by a number of researchers.

Recognition of Arabic continuous speech was addressed by Al-Otaibi (2001). He provided a single-speaker speech dataset for MSA. He also proposed a technique for labeling Arabic speech. He reported a recognition rate for speaker-dependent ASR of 93.78% using his technique. The ASR was built using the Hidden Markov Model (HMM) tool kit (HTK). Hyassat and Abu Zitar (2008) described an Arabic speech recognition system based on Sphinx4. They also proposed an automatic toolkit for building phonetic dictionaries for the Holy Qur'an and standard Arabic language. Three corpuses were developed in this work, namely, the Holy Qura'an corpus HQC-1 of about 18.5 h, the command and control corpus CAC-1 of about 1.5 h, and the Arabic digits corpus ADC of less than 1 h of speech.

A workshop was held in 2002 at John Hopkins University. Kirchhofl et al. (2003) proposed to use Romanization method for transcription of Egyptian dialectic of telephone conversations. Soltau et al. (2007) reported advancements in the IBM system for Arabic speech recognition as part of the continuous effort for the GALE project. The system consists of multiple stages that incorporate both vocalized and non-vocalized Arabic speech model. The system also incorporates a training corpus of 1,800 h of unsupervised Arabic speech. Azmi et al. (2008)

D. AbuZeina and M. Elshafei, *Cross-Word Modeling for Arabic Speech Recognition*, SpringerBriefs in Electrical and Computer Engineering,
DOI 10.1007/978-1-4614-1213-7_2, © Dia AbuZeina 2012

investigated using Arabic syllables for speaker-independent speech recognition system for Arabic spoken digits. The database used for both training and testing consists of 44 Egyptian speakers. In a clean environment, experiments show that the recognition rate obtained using syllables outperformed the rate obtained using monophones, triphones, and words by 2.68%, 1.19%, and 1.79%, respectively. Also in noisy telephone channel, syllables outperformed the rate obtained using monophones, triphones, and words by 2.09%, 1.5%, and 0.9%, respectively. Abdou et al. (2006) described a speech-enabled computer-aided pronunciation learning (CAPL) system. The system was developed for teaching Arabic pronunciations to non-native speakers. The system uses a speech recognizer to detect errors in user recitation. A phoneme duration classification algorithm is implemented to detect recitation errors related to phoneme durations. Performance evaluation using a dataset that includes 6.6% wrong speech segments showed that the system correctly identified the error in 62.4% of pronunciation errors, reported "Repeat Request" for 22.4% of the errors, and made false acceptance of 14.9% of total errors. Khasawneh et al. (2004) compared the polynomial classifier that was applied to isolated-word speaker-independent Arabic speech and dynamic time warping (DTW) recognizer. They concluded that the polynomial classifier produced better recognition performance and much faster testing response than the DTW recognizer. Choi et al. (2008) presented recent improvements to their English/Iraqi Arabic speech-to-speech translation system. The presented system-wide improvements include user interface (UI), dialog manager, ASR, and machine translation (MT) components. Rambow et al. (2006) addressed the problem of parsing transcribed spoken Arabic. They examined three different approaches: sentence transduction, treebank transduction, and grammar transduction. Overall, grammar transduction outperformed the other two approaches. Parsing can be used to check the speech recognizer n-best hypothesis to rescore them according to most syntactically accurate one. Nofal et al. (2004) demonstrated a design and implementation of stochastic-based new acoustic models suitable for use with a command and control system speech recognition system for the Arabic language. Park et al. (2009) explored the training and adaptation of multilayer perceptron (MLP) features in Arabic ASRs. Three schemes had been investigated. First, the use of MLP features to incorporate short-vowel information into the graphemic system. Second, a rapid training approach for use with the perceptual linear predictive (PLP) + MLP system was described. Finally, the use of linear input networks (LIN) adaptation as an alternative to the usual HMM-based linear adaptation was demonstrated. Shoaib et al. (2003) presented a novel approach to develop a robust Arabic speech recognition system based on a hybrid set of speech features. This hybrid set consists of intensity contours and formant frequencies. Imai et al. (1995) presented a new method for automatic generation of speaker-dependent phonological rules in order to decrease recognition errors caused by pronunciation variability dependent on speakers. Choueiter et al. (2006) concentrated our efforts on MSA, where they built morpheme-based LMs and studied their effect on the OOV rate as well as the word error rate (WER). Bourouba et al. (2006) presented a new HMM/support vectors machine (SVM) (*k*-nearest neighbor) for recognition of isolated spoken words. Sagheer et al. (2005) presented a novel visual speech features representation system. They used it to comprise a complete lip-reading system. Taha et al. (2007) demonstrated a novel agent-based design for Arabic speech recognition. They defined the Arabic speech recognition as a Multi-agent System where each agent has a specific goal and deals with that goal only. Elmisery et al. (2003) implemented a pattern matching algorithm based on HMM using field programmable gate array (FPGA). The proposed approach was used for isolated Arabic word recognition and achieved accuracy comparable with the powerful classical recognition system. Mokhtar and El-Abddin (1996) represented the techniques and algorithms used to model the acoustic-phonetic structure of Arabic speech recognition using HMMs. Gales et al. (2007) described the development of a phonetic system for Arabic speech recognition. A number of issues involved with building these systems had been discussed, such as the pronunciation variation problem. Bahi and Sellami (2001) presented experiments performed to recognize isolated Arabic words. Their recognition system was based on a combination of the vector quantization technique at the acoustic level and Markovian modeling.

A number of researchers investigated the use of neural networks for Arabic phonemes and digits recognition (El-Ramly et al. 2002; Bahi and Sellami 2003; Shoaib et al. 2003). For example, El-Ramly et al. (2002) studied recognition of Arabic phonemes using an Artificial Neural Network. Alimi and Ben Jemaa (2002) proposed the use of a Fuzzy Neural Network for recognition of isolated words. Bahi and Sellami (2003) investigated a hybrid of neural networks and HMMs for NN/HMM for speech recognition. Alotaibi (2004) reported achieving high-performance Arabic digits recognition using recurrent networks. Essa et al. (2008) proposed different combined classifier architectures based on Neural Networks by varying the initial weights, architecture, type, and training data to recognize Arabic isolated words. Emami and Mangu (2007) studied the use of neural network language models (NNLMs) for Arabic broadcast news and broadcast conversations speech recognition.

Alghamdi et al. (2009) developed an Arabic broadcast news transcription system. They used a corpus of 7.0 h for training and 0.5 h for testing. They achieved a WER of 8.61%. The WER obtained ranged from 14.9 to 25.1% for different types and sizes of test data. Satori et al. (2007) used Sphinx tools for Arabic speech recognition. They demonstrated the use of the tools for recognition of isolated Arabic digits. The data were recorded from six speakers. They achieved a digits recognition accuracy of 86.66%. Lamel et al. (2009) described the incremental improvements to a system for the automatic transcription of broadcast data in Arabic, highlighting techniques developed to deal with specificities (no diacritics, dialectal variants, and lexical variety) of the Arabic language. Afify et al. (2005) compared grapheme-based recognition system with explicitly modeling short vowels. They found that short vowels modeling improves recognition performance. Billa et al. (2002) described the development of audio indexing system for broadcast news in Arabic. Key issues addressed in this work revolve around the three major components of the audio indexing system: automatic speech recognition, speaker identification, and named entity identification.

Messaoudi et al. (2006) demonstrated that by building a very large vocalized vocabulary and by using a language model including a vocalized component, the WER could be significantly reduced. Elmahdy et al. (2009) used acoustic models trained with large MSA news broadcast speech corpus to work as multilingual or multi-accent models to decode colloquial Arabic. Vergyri et al. (2004) showed that the use of morphology-based language models at different stages in a large-vocabulary continuous speech recognition (LVCSR) system for Arabic leads to WER reductions. To deal with the huge lexical variety, Xiang et al. (2006) concentrated on the transcription of Arabic broadcast news by utilizing morphological decomposition in both acoustic and language modeling in their system. Selouani and Alotaibi (2011) presented Genetic Algorithms to adapt HMMs for non-native speech in a large-vocabulary speech recognition system of MSA. Saon et al. (2010) described the Arabic broadcast transcription system fielded by IBM in the GALE project. Key advances include improved discriminative training, the use of subspace Gaussian mixture models (SGMM), neural network acoustic features, variable frame rate decoding, training data partitioning experiments, unpruned n-gram language models, and NNLMs. These advances were instrumental in achieving a WER of 8.9% on the evaluation test set. Kuo et al. (2010) studied various syntactic and morphological context features incorporated in an NNLM for Arabic speech recognition.

2.2 Arabic Speech Recognition Challenges

Arabic speech recognition faces many challenges. For example, Arabic has short vowels which are usually ignored in text. Therefore, more confusion will be added to the ASR decoder. Additionally, Arabic has many dialects where words are pronounced differently. Elmahdy et al. (2009) summarized the main problems in Arabic speech recognition, which include Arabic phonetics, diacritization problem, grapheme-to-phoneme relation, and morphological complexity. Diacritization is represented by different possible diacritizations of a particular word. As modern Arabic is usually written in non-diacritized scripts, lots of ambiguities for pronunciations and meanings are introduced. Elmahdy et al. (2009) also showed that grapheme-to-phoneme relation is only true for diacritized Arabic script. Arabic morphological complexity is demonstrated by the large number of affixes (prefixes, infixes, and suffixes) that can be added to the three consonant radicals to form patterns. Farghaly and Shaalan (2009) provided a comprehensive study of Arabic language challenges and solutions. Lamel et al. (2009) presented a number of challenges for Arabic speech recognition such as no diacritics, dialectal variants, and very large lexical variety. Alotaibi et al. (2008) introduced foreign-accented Arabic speech as a challenging task in speech recognition. A workshop was held in 2002 at John Hopkins University to define and address the challenges in developing a speech recognition system using Egyptian dialectic Arabic for telephone conversations. They proposed to use Romanization method for transcription of the speech corpus (Kirchhofl et al. 2003). Abushariah et al. (2010) reported the design, implementation, and evaluation of a research work for developing a high-performance natural speaker-independent Arabic continuous speech recognition system. Muhammad et al. (2011) evaluated conventional ASR system for six different types of voice disorder patients speaking Arabic digits. MFCC and Gaussian mixture model (GMM)/HMM are used as features and classifier, respectively. Recognition result is analyzed for types of diseases. Billa et al. (2002) discussed a number of research issues for Arabic speech recognition, e.g., absence of short vowels in written text and the presence of compound words that are formed by the concatenation of certain conjunctions, prepositions, articles, and pronouns, as prefixes and suffixes to the word stem.

References

Abdou SM, Hamid SE, Rashwan M, Samir A, Abd-Elhamid O, Shahin M, Naz W (2006) Computer aided pronunciation learning system using speech recognition techniques, NTERSPEECH 2006, ICSLP, pp 249–252

Abushariah MAM, Ainon RN et al (2010) Natural speaker-independent Arabic speech recognition system based on Hidden Markov Models using Sphinx tools. 2010 international conference on computer and communication engineering (ICCCE)

Afify M, Nguyen L, Xiang B, Abdou S, Makhoul J. Recent progress in Arabic broadcast news transcription at BBN. In: Proceedings of INTERSPEECH. 2005, pp 1637–1640

Algamdi M (2003) KACST Arabic phonetics database. The fifteenth international congress of phonetics science, Barcelona, pp 3109–3112

Alghamdi M (2000) Arabic phonetics. Attaoobah, Riyadh

Alghamdi M, Elshafei M, Almuhtasib H (2002) Speech units for Arabic text-to-speech. The fourth workshop on computer and information sciences, pp 199–212

Alghamdi M, Elshafei M, Almuhtasib H (2009) Arabic broadcast news transcription system. Int J Speech Tech 10:183–195

Al-Ghamdi M, Elshafei M, Al-Muhtaseb H (2003) An experimental Arabic text-to-speech system. Final report, King Abudaziz City of Science and Technology

Alimi AM, Ben Jemaa M (2002) Beta fuzzy neural network application in recognition of spoken isolated Arabic words. Int J Contr Intell Syst 30(2), Special issue on speech processing techniques and applications

Alotaibi YA (2004) Spoken Arabic digits recognizer using recurrent neural networks. In: Proceedings of the fourth IEEE international symposium on signal processing and information technology, pp 195–199

Al-Otaibi F (2001) speaker-dependant continuous Arabic speech recognition. M.Sc. thesis, King Saud University

Alotaibi Y, Selouani S, O'Shaughnessy D (2008) Experiments on automatic recognition of nonnative Arabic speech. EURASIP J Audio Speech Music Process: 9 pages. doi:10.1155/2008/679831, Article ID 679831

Azmi M, Tolba H, Mahdy S, Fashal M (2008) Syllable-based automatic Arabic speech recognition in noisy-telephone channel. In: WSEAS transactions on signal processing proceedings, World Scientific and Engineering Academy and Society (WSEAS), vol 4, issue 4, pp 211–220

Bahi H, Sellami M (2001) Combination of vector quantization and hidden Markov models for Arabic speech recognition. ACS/IEEE international conference on computer systems and applications, 2001

Bahi H, Sellami M (2003) A hybrid approach for Arabic speech recognition. ACS/IEEE international conference on computer systems and applications, 14–18 July 2003

Billa J, Noamany M et al (2002) Audio indexing of Arabic broadcast news. 2002 IEEE international conference on acoustics, speech, and signal processing (ICASSP)

Bourouba H, Djemili R et al (2006) New hybrid system (supervised classifier/HMM) for isolated Arabic speech recognition. 2nd Information and Communication Technologies, 2006. ICTTA'06

Choi F, Tsakalidis S et al (2008) Recent improvements in BBN's English/Iraqi speech-to-speech translation system. IEEE Spoken language technology workshop, 2008. SLT 2008

Choueiter G, Povey D et al (2006) Morpheme-based language modeling for Arabic LVCSR. 2006 IEEE international conference on acoustics, speech and signal processing. ICASSP 2006 proceedings

Elmahdy M, Gruhn R et al (2009) Modern standard Arabic based multilingual approach for dialectal Arabic speech recognition. In: Eighth international symposium on natural language processing, 2009. SNLP'09

Elmisery FA, Khalil AH et al (2003) A FPGA-based HMM for a discrete Arabic speech recognition system. In: Proceedings of the 15th international conference on microelectronics, 2003. ICM 2003

El-Ramly SH, Abdel-Kader NS, El-Adawi R (2002) Neural networks used for speech recognition. In: Proceedings of the nineteenth national radio science conference (NRSC 2002), March 2002, pp 200–207

Elshafei MA (1991) Toward an Arabic text-to-speech system. Arab J Sci Eng 16(4B):565–583

Elshafei M, Almuhtasib H, Alghamdi M (2002) Techniques for high quality text-to-speech. Inform Sci 140(3–4):255–267

Elshafei M, Al-Muhtaseb H, Alghamdi M (2006) Statistical methods for automatic diacritization of Arabic text. In: Proceedings of 18th national computer conference NCC'18, Riyadh, March 26–29, 2006

Elshafei M, Ali M, Al-Muhtaseb H, Al-Ghamdi M (2007) Automatic segmentation of Arabic speech. Workshop on information technology and Islamic sciences, Imam Mohammad Ben Saud University, Riyadh, March 2007

Emami A, Mangu L (2007) Empirical study of neural network language models for Arabic speech recognition. IEEE workshop on automatic speech recognition and understanding, 2007. ASRU

Essa EM, Tolba AS et al (2008) A comparison of combined classifier architectures for Arabic speech recognition. International conference on computer engineering and systems, 2008. ICCES 2008

Farghaly A, Shaalan K (2009) Arabic natural language processing: challenges and solutions. ACM Trans Asian Lang Inform Process 8(4):1–22

Gales MJF, Diehl F et al (2007) Development of a phonetic system for large vocabulary Arabic speech recognition. IEEE workshop on automatic speech recognition and understanding, 2007. ASRU

Hyassat H, Abu Zitar R (2008) Arabic speech recognition using SPHINX engine. Int J Speech Tech 9(3–4):133–150

Imai T, Ando A et al (1995) A new method for automatic generation of speaker-dependent phonological rules. 1995 international conference on acoustics, speech, and signal processing, 1995. ICASSP-95

Khasawneh M, Assaleh K et al (2004) The application of polynomial discriminant function classifiers to isolated Arabic speech recognition. In: Proceedings of the IEEE international joint conference on neural networks, 2004

Kirchhofl K, Bilmes J, Das S, Duta N, Egan M, Ji G, He F, Henderson J, Liu D, Noamany M, Schoner P, Schwartz R, Vergyri D (2003) Novel approaches to Arabic speech recognition: report from the 2002 John-Hopkins summer workshop, ICASSP 2003, pp I344–I347

Kuo HJ, Mangu L et al (2010) Morphological and syntactic features for Arabic speech recognition. 2010 IEEE international conference on acoustics speech and signal processing (ICASSP)

Lamel L, Messaoudi A et al (2009) Automatic speech-to-text transcription in Arabic. ACM Trans Asian Lang Inform Process 8(4):1–18

Messaoudi A, Gauvain JL et al (2006) Arabic broadcast news transcription using a one million word vocalized vocabulary. 2006 IEEE international conference on acoustics, speech and signal processing, 2006. ICASSP 2006 proceedings

Mokhtar MA, El-Abddin AZ (1996) A model for the acoustic phonetic structure of Arabic language using a single ergodic hidden Markov model. In: Proceedings of the fourth international conference on spoken language, 1996. ICSLP 96

Muhammad G, AlMalki K et al (2011) Automatic Arabic digit speech recognition and formant analysis for voicing disordered people. 2011 IEEE symposium on computers and informatics (ISCI)

Nofal M, Abdel Reheem E et al (2004) The development of acoustic models for command and control Arabic speech recognition system. 2004 international conference on electrical, electronic and computer engineering, 2004. ICEEC'04

Park J, Diehl F et al (2009) Training and adapting MLP features for Arabic speech recognition. IEEE international conference on acoustics, speech and signal processing, 2009. ICASSP 2009

Rambow O et al (2006) Parsing Arabic dialects, final report version 1, Johns Hopkins summer workshop 2005

Sagheer A, Tsuruta N et al (2005) Hyper column model vs. fast DCT for feature extraction in visual Arabic speech recognition. In: Proceedings of the fifth IEEE international symposium on signal processing and information technology, 2005

Saon G, Soltau H et al (2010) The IBM 2008 GALE Arabic speech transcription system. 2010 IEEE international conference on acoustics speech and signal processing (ICASSP)

Satori H, Harti M, Chenfour N (2007) Introduction to Arabic speech recognition using CMU Sphinx system. Information and communication technologies international symposium proceeding ICTIS07, 2007

Selouani S-A, Alotaibi YA (2011) Adaptation of foreign accented speakers in native Arabic ASR systems. Appl Comput Informat 9(1):1–10

Shoaib M, Rasheed F, Akhtar J, Awais M, Masud S, Shamail S (2003) A novel approach to increase the robustness of speaker independent Arabic speech recognition. 7th international multi topic conference, 2003. INMIC 2003. 8–9 Dec 2003, pp 371–376

Soltau H, Saon G et al (2007) The IBM 2006 Gale Arabic ASR system. IEEE international conference on acoustics, speech and signal processing, 2007. ICASSP 2007

Taha M, Helmy T et al (2007) Multi-agent based Arabic speech recognition. 2007 IEEE/WIC/ACM international conferences on web intelligence and intelligent agent technology workshops

Vergyri D, Kirchhoff K, Duh K, Stolcke A (2004) Morphology-based language modeling for Arabic speech recognition. International conference on speech and language processing. Jeju Island, pp 1252–1255

Xiang B, Nguyen K, Nguyen L, Schwartz R, Makhoul J (2006) Morphological ecomposition for Arabic broadcast news transcription. In: Proceedings of ICASSP, vol I. Toulouse, pp 1089–1092

Chapter 3
The Baseline System

This chapter presents the main components of the baseline system that was used to test the knowledge-based proposed method. The Arabic speech corpus, Arabic phoneme set, Arabic language model, and Arabic pronunciation dictionary are described. The chapter also provides the details of how to build each one of these Arabic automatic speech recognition (AASR) components.

3.1 Arabic Speech Corpus

We utilized a large vocabulary, speaker-independent, natural Arabic continuous speech recognition system developed at King Fahd University of Petroleum and Minerals (KFUPM). The AASR system is based on Sphinx-III CMU ASR system. The baseline system uses three emitting states of HMM for triphone-based acoustic models. The state probability distribution uses a continuous density of eight Gaussian mixture distributions. The baseline system is trained using audio files recorded from several TV news channels at a sampling rate of 16 k samples per seconds. Our corpus includes 249 business/economics and sports stories (144 by male speakers, 105 by female speakers), summing up to 5.4 h of speech. The 5.4 h were split into 4,572 files with an average file length of 4.5 s. The length of wave files ranges from 0.8 to 15.6 s. An additional 0.1-s silence period is added to the beginning and end of each file. Although care was taken to exclude recordings with background music or excessive noise, some of the files still contain background noise such as low level or fainting music; environmental noise such as that of a reporter in an open area, e.g., a stadium or a stock market; and low level overlapping foreign speech, occurring when a reporter is translating foreign statements. The 4,572 wave files were completely transcribed with fully diacritized text.

The transcription is meant to reflect the way the speaker has uttered the words, even if they were grammatically wrong. It is a common practice in MSA and most Arabic dialects to drop the vowels at the end of words; this situation is

D. AbuZeina and M. Elshafei, *Cross-Word Modeling for Arabic Speech Recognition*,
SpringerBriefs in Electrical and Computer Engineering,
DOI 10.1007/978-1-4614-1213-7_3,

represented in the transcription by either using a silence mark (Sukun or unvoweled) or dropping the vowel, which is considered equivalent to the silence mark. The transcription file contains 39,217 words. The vocabulary list contains 14,234 words. The baseline WER is 12.21%.

3.2 Arabic Phoneme Set

Before proceeding in discussing the Arabic phoneme set, it would be easier for the reader if we start first by providing a Romanization of the Arabic letters and diacritical marks as shown in Table 3.1. We used Ryding (2005) as a reference. The short vowels Fatha, Damma, and Kasra are represented using a, u, and i, respectively.

A phoneme is the basic unit of speech that is used in ASR systems. Table 3.2 shows the listing of the Arabic phoneme set used in the training, and the corresponding phoneme symbols. The table also shows illustrative examples of vowel usage. This phoneme set is chosen based on the previous experience with Arabic text-to-speech systems (Elshafei 1991; Alghamdi et al. 2004; Elshafei et al. 2002), and the corresponding phoneme set that is successfully used in the CMU English Pronunciation Dictionary (2011). Although we verified the Arabic phoneme set and found it to be good enough, we believe that this set is far from being optimal, and further work is needed to derive an optimized phoneme set for AASR and possibly for bilingual speech recognition applications.

The regular *Arabic short vowels* are /AE/, /IH/, and /UH/ corresponding to the Arabic diacritical marks Fatha, Damma, and Kasra, respectively. The /AA/ is the pharyngealized allophone of /AE/, which appears after an emphatic letter. Similarly, the /IX/ and /UX/ are the pharyngealized allophones of /IH/ and /UH/, respectively. When /AE/ appears before an emphatic letter, its allophone /AH/ is used instead.

The regular *Arabic long vowel* allophones are /AE:/ /IY/ and /UW/, respectively. The length of a long vowel is normally equal to two short vowels. The allophones / AY/ and /AW/ are actually two vowel sounds in which the articulators move from one post to another. These vowels are called diphthongs. The allophone

Table 3.1 Arabic–Roman letters mapping table

Arabic	Roman	Arabic	Roman	Arabic	Roman	Arabic	Roman
ء (hamza)	’	د (daal)	d	ض (Daad)	D	ك (kaaf)	k
ب (baa’)	b	ذ (dhaal)	dh	ط (Taa’)	T	ل (laam)	l
ت (taa’)	t	ر (raa’)	r	ظ (Zaa’)	Z	م (miim)	m
ث (thaa’)	th	ز (zaay)	z	ع (‘ayn)	‘	ن (nuun)	n
ج (jiim)	j	س (siin)	s	غ (ghayn)	gh	ه (haa’)	h
ح (Haa’)	H	ش (shiin)	sh	ف (faa’)	f	و (waaw)	w or u
خ (khaa’)	kh	ص (Saad)	S	ق (qaaf)	q	ي (yaa’)	y or ii

Table 3.2 The phonemes set used in the training

Phoneme	Example ◄ Letter	Phoneme	Letter
/AE/	بَ◄ ـَ Fatha	/DH/	ذ (dhaal)
/AE:/	بَاب◄ ـَا	/R/	ر (raa’)
/AA/	خَ◄ ـَ Hard Fatah	/Z/	ز (zaay)
/AH/	دَ◄ ـَ Soft Fatah	/S/	س (siin)
/UH/	بُ◄ ـُ Damma	/SH/	ش (shiin)
/UW/	دُون◄ ـُو	/SS/	ص (Saad)
/UX/	غُصن◄ ـُ	/DD/	ض (Daad)
/IH/	بِنت◄ ـِ Kasra	/TT/	ط (Taa’)
/IY/	فِيل◄ ـِي	/DH2/	ظ (Zaa’)
/IX/	صِنف◄ ـِ	/AI/	ع (‘ayn)
/AW/	لوم◄ ـُو	/GH/	غ (ghayn)
/AY/	ضَيف◄ ـَي	/F/	ف (faa’)
/E/	ء (hamza)	/Q/	ق (qaaf)
/B/	ب (baa’)	/K/	ك (kaaf)
/T/	ت (taa’)	/L/	ل (laam)
/TH/	ث (thaa’)	/M/	م (miim)
/JH/	ج (jiim)	/N/	ن (nuun)
/HH/	ح (Haa’)	/H/	ه (haa’)
/KH/	خ (khaa’)	/W/	و (waaw)
/D/	د (daal)	/Y/	ي (yaa’)

/AY/ appears when a Fatha comes before an undiacritized yaa’. Similarly, /AW/ appears when a Fatha comes before an undiacritized waaw.

The *Arabic voiced stops* phonemes /B/ and /D/ are similar to their English counterparts. /DD/ corresponds to the sound of the Arabic Daad letter.

The *Arabic voiceless stops* /T/ and /K/ are basically similar to their English counterparts, while /TT/ is the emphatic version of /T/.

The sound of the Arabic emphatic letter qaaf is represented by the phone /Q/. The Hamza plosive sound is represented by the phone /E/.

The *voiceless fricatives* are produced with no vibration of the voice cords. The sound is produced by the turbulence flow of air through a constriction. The Arabic voiceless fricatives /F/, /S/, /TH/, /SH/, and /H/ are basically similar to their English twins. In addition, the Arabic phones /SS/, /HH/, and /KH/ are the sounds of the Arabic letters Saad, Haa’, and khaa’, respectively.

Voiced fricatives are generated with simultaneous vibration of the vocal cords. The Arabic voiced fricative phones are /AI/, /GH/, /Z/, and /DH/ corresponding to the sound of the Arabic letters ‘ayn, ghayn, zaay, and dhaal. /DH2/ is the emphatic version of /DH/.

The *Arabic affricative sound* /JH/ is similar to the corresponding one in English. The Arabic resonants are similar to the English resonant phones. These are /Y/ for yaa', /W/ for waaw, /L/ for laam, and /R/ for raa'.

The *Nasal phonemes* include /M/ and /N/ which are similar to the corresponding English sounds.

3.3 Arabic Pronunciation Dictionary

Pronunciation dictionaries are essential components of ASRs. They contain the phonetic transcriptions of all the vocabulary in the target domain of the conversation. A phonetic transcription is a sequence of phonemes that describes how the corresponding word should be pronounced. Ali et al. (2009) developed a software tool to generate phonetic dictionaries for Arabic texts using Arabic pronunciation rules. We utilized this tool to generate the enhanced dictionary (i.e., after modeling cross-word problem). The tool developed by Ali et al. (2009) takes care of some of within-word variations such as the context in which the words are uttered. For example, Hamzat Al-Wasl (ا) at the beginning of the word, the Ta'al marbouta (ة) at the end of the word, and words and letters that have multiple pronunciations due to dialect issues. They also defined a set of rules based on regular expressions to define the phonemic definition of words. The tool scans a word letter by letter, and if the conditions of a rule for a specific letter are satisfied, then the replacement for that letter is added to a tree structure that represents all the possible pronunciations for that word.

Each rule has the following structure:
LETTER:

(pre_condition) · (post_condition) -> replacement,

where LETTER represents the current letter in the word, pre_condition and post_condition are regular expressions that represent other letters surrounding the current letter, and replacement is the replacement phoneme or phonemes. The baseline dictionary contains 14,234 words (without variants) and 23,840 words (with within-word variants). A sample from the developed pronunciation dictionary is listed below. This example shows the within-word variants of (أدِنبَرَة <> 'dinbara), in the baseline dictionary:

```
أدنبرة      E AE D IH M B R AA H (default)
أدنبرة(2)   E AE D IH M B R AA T
أدنبرة(3)   E AE D IH N B R AA H
أدنبرة(4)   E AE D IH N B R AA T
```

The defined rules are provided for each Arabic letter available in the Unicode listing (45 letters). Each rule tries to match certain conditions on the context of the letter and provides a replacement from the phoneme list. Replacements can be one or more phonemes. Some letters do not have an effect on pronunciation, or depending on context, they might not be pronounced; in this case, the replacement will be empty.

For the rule format:

(pre_condition) · (post_condition) -> replacement,

The left-hand side of the rule is a PERL-like regular expression with the following definitions.

Each letter in the Arabic alphabet is referenced by its name as defined in the Unicode standard. The dot (·) in the middle marks the current position (which is also the current letter) in the word. Multiple classes are defined to simplify the rules' syntax. Each class is referenced by its symbol (L, D, S, etc.) surrounded by angle brackets (< >).

The classes are as follows:

<L>: All Arabic consonants.
<D>: Diacritic marks [Fathatan (ً), Dammatan (ٌ), Kasratan (ٍ), Fatha (َ), Damma (ُ), Kasra (ِ), Shadda (ّ), and Sukun (ْ)].
<S>: Word Start.
<T>: Word End.
<SH>: Shamsi Letters (taa', thaa', daal, dhaal, raa', zaay, siin, shiin, Saad, Daad, Taa', Zaa', laam, and nuun).
<V>: Vowels (Fatha, Damma, Kasra, and Shadda).
<VA>: Vowels without Shadda (Fatha, Damma, and Kasra).
<P>: Prefix letters (waaw, baa', faa', kaaf, and laam).
<E>: Emphatic letters (Tah, Saad, Daad, and Zaa').
<PH>: Pharyngeal letters (qaaf, ghayn, khaa', and raa').

The precondition has one of the following formats: (?<=pattern): context before the current position matches the pattern. (?<!pattern): context before the current position does not match the pattern. In the same way, the post-condition has one of the following formats: (?=pattern): context after the current position matches the pattern. (?!pattern): context after the current position does not match the pattern.

Patterns use the following operators to define expressions:

Alternation: a vertical bar (|) is used to separate alternatives. Grouping: parentheses ( ) are used to define groups that determine scope and precedence of the operators and build complex expressions. Optional matching: a question mark (?) is used to mark parts of the expression that may or may not exist.

The right-hand side of the rule defines the replacement, which can either be a phoneme or a sequence of phonemes from the phoneme list, or the letter might not have a matching phoneme and will be omitted from pronunciation. This case is marked with an asterisk (*) on the right-hand side.

We define a rule set that covers all possible Arabic letters that are used in typing. Many of the rules are straight forward; they match the Arabic letters to their corresponding phonemes as explained in Table 3.1. Vowels require more elaborate rules to cover all possibilities. Special attention is required for nasalized consonants (Meem and Nuun) and a few more exceptions that will be explained in the following sections.

3.3.1 Consonants

All Arabic consonants have a directly matching phoneme. However, some letters (thaa', jiim, dhaal, Zaa', and qaaf) have multiple possible pronunciations that are due to dialectical differences.

A nuun followed by a baa' is usually converted to a miim. This rule is optional and the speaker might not follow.

nuun:

.(?= baa')-> M
.-> N

If the letter daal is followed by a voweled taa', then it is omitted in pronunciation. A similar case applies to the letter daad.

daal:

.(?= taa'<V>)-> *
.(?! taa' <V>)-> D

3.3.2 The Letter Laam

The letter laam has a set of complex rules when it comes in a combination known as Al-Alta'rif (ـال), which is Alef (ا) and laam (ل). If this combination is followed by a letter from the Shamsi group, then the laam is not pronounced. This rule, however, is not mandatory.

laam:

(?<=(<P><V>)?ALEF FATHA?).(?= <SH>)-> *
.-> L

3.3.3 Semivowels

The letter waaw is sometimes treated as semiconsonants /W/ or /AW/, and other times it is treated as a long vowel, depending on its context. If the letter waaw is not voweled and is proceeded by a Damma, then it is considered to be a long vowel.

The case of the semivowel /AW/ is similar to that of the long vowel, except it is then preceded by a Fatha. In this case, the waaw is omitted. The insertion of the /AW/ phoneme is handled by the Fatha rules as it will follow shortly.

In the rest of the cases, the waaw is converted to the semivowel /W/.

waaw:

(?<=(FATHA|DAMMA)).(?!<V>)-> *

(?<=(FATHA|DAMMA)).(?=<V>)-> W

(?<!(FATHA|DAMMA)).-> W

The letter yaa' follows a similar pattern.

3.3.4 *Tanween*

The rules for this group differentiate between the emphatic and/or the pharyngeal versions of the vowels. Each rule appends an /N/ sound to the pronunciation.

FATHATAN:

(?<!<E>|<PH>).-> AE N

(?<=<E>).-> AH N

(?<=<PH>).-> AA N

DAMMATAN:

(?<!<PH>).-> UH N

(?<=<PH>).-> UX N

KASRATAN:

(?<!<PH>).-> IH N

(?<=<PH>).-> IX N

3.3.5 *Vowels*

Vowels have many versions. They are either short or long vowels. Both short and long vowels also can be normal ones, or either emphatic or pharyngeal, depending on the surrounding letters. We developed rules that take care of all these situations. For instance, the rules for Fatha are as follows:

FATHA:

1. (?<!<E>|<PH>).(?!ALEF|((waaw | yaa')(<L>|<T>)))-> AE
2. (?<!<E>|<PH>).(?=ALEF)-> AE:
3. (?<=ALEF_WITH_MADDA_ABOVE).-> AE:

4. (?<=<E>).(?!ALEF|((waaw | yaa')(<L>|<T>)))-> AH
5. (?<=<E>).(?=ALEF)-> AH:
6. (?<=<PH>).(?!ALEF|((waaw | yaa')(<L>|<T>)))-> AA
7. (?<=<PH>).(?=ALEF)-> AA:
8. .(?=WAW (<L>|<T>))-> AW
9. .(?=YEH (<L>|<T>))-> AY

The first two rules are responsible for the long and short versions of the normal vowels. The third rule is also for long vowels where the Fatha is followed by an Alef with Madda above.

Rules 4–7 are for the emphatic and pharyngeal versions of the vowel.

Rules 8 and 9 take care of the semivowels /AW/ and /AY/, as mentioned in the rules for the Waw and yaa'.

Rules for Damma and Kasra follow the same logic as that for the FATHA.

3.4 Arabic Language Model

We used the CMU language toolkit (Open Source Toolkit for Speech Recognition 2011) to build a statistical language model from the transcription of the full diacritized transcription of 5.4 h of the audio. Table 3.3 shows the total count of 1-grams, 2-grams, and 3-grams of the Arabic baseline language model with examples.

Table 3.3 n-Grams in the baseline system

n-grams Type	n-grams count	Examples
1-grams	14234	أضحَو <> 'DHaw أضعَافِ <> 'D 'aafi أضَحَت <> 'DHat
2-grams	32813	المَجلِس الإتّحَادِيّ <> almajlis al'tHaadyi المَجلِس العَالَمِيّ <> almajlis al'aalamyi المَجلِس تَعَامُلاتِهَا <> almajlis t'amulatiha
3-grams	37771	المَعنِيَة وَالتَّأكِيدِ عَلَى <> alma'niya walta'kiid 'ala المَعنِيَّة خَمسَةِ مِليَارَات <> alma'niya khmsh mlyarat المَعنِيَّة فِي المَطَار <> alma'niya fy almatar

References

Ali M, Elshafei M, Alghamdi M, Almuhtaseb H, Alnajjar A (2009) Arabic phonetic dictionaries for speech recognition. J Inform Tech Res 2(4):67–80

Alghamdi M, Almuhtasib H, Elshafei M (2004) Arabic phonological rules. King Saud Univ J Comput Sci Inf 16:1–25

Elshafei MA (1991) Toward an Arabic text-to-speech system. Arab J Sci Eng 16(4B):565–583

Elshafei M, Almuhtasib H, Alghamdi M (2002) Techniques for high quality text-to-speech. Inf Sci 140(3–4):255–267

Open Source Toolkit for Speech Recognition (2011) http://cmusphinx.sourceforge.net/wiki/download/. Accessed 1 Sep 2011

Ryding KC (2005) A reference grammar of modern standard Arabic (Reference grammars). Cambridge University Press, Cambridge

The CMU Pronunciation Dictionary (2011) http://www.speech.cs.cmu.edu/cgi-bin/cmudict. Accessed 1 Sept 2011

Chapter 4
Cross-Word Pronunciation Variations

This chapter presents the cross-word problem of the Arabic language. It also includes the main sources of this problem: Idgham (merging), Iqlaab (changing), Hamzat Al-Wasl deleting, and merging of two consecutive unvoweled letters. Illustrative examples of the cross-word problem are also provided.

4.1 Introduction

The main goal of automatic speech recognition systems (ASRs) is to enable people to communicate more naturally and effectively. But this ultimate dream faces many obstacles such as variability in speaking styles and pronunciation variations. As we mentioned in Chap. 1, Benzeghiba et al. (2007) presented the speech variability sources: foreign and regional accents, speaker physiology, spontaneous speech, rate of speech, children speech, emotional state, noises, new words, and more. Accordingly, handling these obstacles is a major requirement to enhance ASR performance.

In speech recognition, pronunciation variation causes recognition errors in the form of insertions, deletions, or substitutions of phoneme(s) relative to the phonemic transcription in the pronunciation dictionary. Pronunciation variations which reduce recognition performance (McAllester et al. 1998) occur in continuous speech in two types: cross-word variation and within-word variation. Within-word variations cause alternative pronunciation(s) within words. In contrast, a cross-word variation occurs in continuous speech in which a sequence of words forms a compound word that should be treated as one entity. Hofmann et al. (2010) demonstrated that conversational speech poses high challenge to nowadays' ASR as people tend to combine or even miss words out.

The pronunciation variations are often modeled using two approaches: knowledge based and data driven. The knowledge-based approach depends on linguistic criteria that have been developed over decades. These criteria are presented as phonetic rules that can be used to find the possible pronunciation alternative(s) for word utterances. On the contrary, data-driven methods depend solely on the training pronunciation

D. AbuZeina and M. Elshafei, *Cross-Word Modeling for Arabic Speech Recognition*,
SpringerBriefs in Electrical and Computer Engineering,
DOI 10.1007/978-1-4614-1213-7_4,

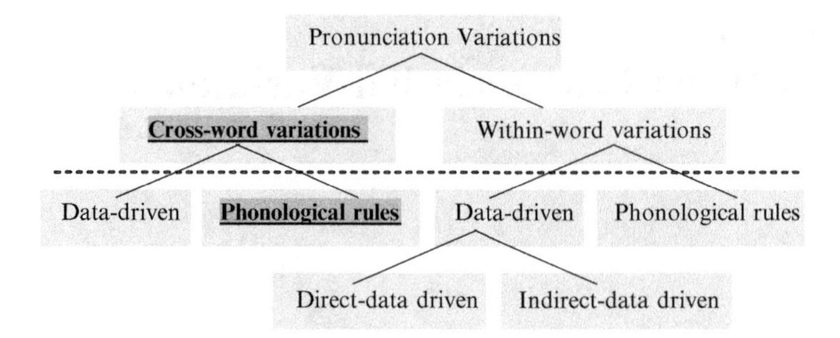

Fig. 4.1 Pronunciation variations and modeling techniques

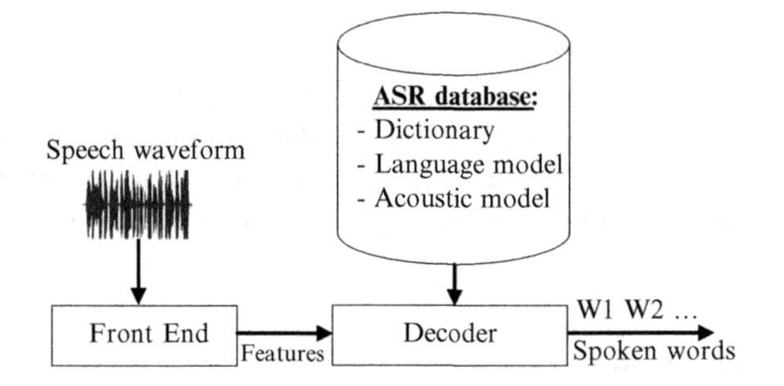

Fig. 4.2 ASR knowledge sources

corpus to find the pronunciation variants (direct data driven) or transformation rules (indirect data driven). That is, the direct data-driven approach distils variants, while the indirect data-driven approach distils rules that are used to find variants. As pros and cons of both approaches, the knowledge-based approach is not exhaustive; not all of the variations that occur in continuous speech can be described, whereas obtaining reliable information using the data-driven approach is extremely difficult (Wester 2003). However, Amdal and Fosler-Lussier (2003) mentioned that there is a growing interest in data-driven methods over the knowledge-based methods due to the lack of domains expertise. Wester and Fosler-Lussier (2000) compared between knowledge-based and data-driven approaches. The comparison showed that the latter leads to significant improvement more than knowledge-based methods which lead to a small improvement in recognition accuracy. Figure 4.1 displays the two types of pronunciation variations and the modeling techniques.

In the figure, the underlined bold text (i.e., modeling cross-word variations using phonological rules) shows the subject research areas of this book. The figure also distinguishes between the types of variations and the modeling techniques by a dashed line: the variation types are above the dashed line, whereas the modeling techniques are under the line.

Modeling cross-word problem requires sufficient knowledge in ASR components. An ASR contains three knowledge sources: acoustic models, dictionary, and language model. Figure 4.2 shows these independent knowledge sources which are collectively

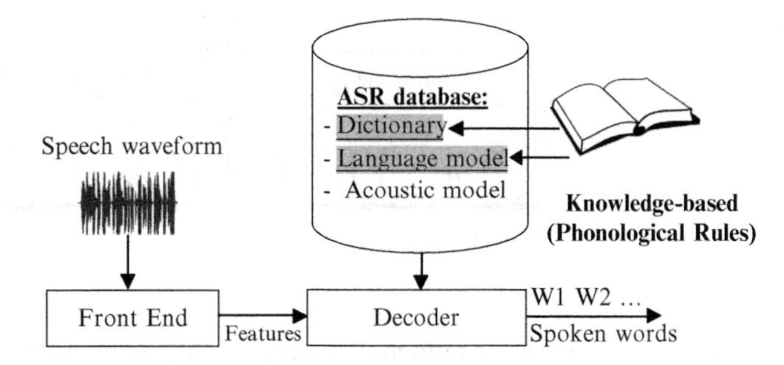

Fig. 4.3 Cross-word adaptation levels

called ASR database. The ASR database is where the pronunciation variations need to be modeled. Despite that the dictionary adaptation dominates the research effort in pronunciation variation modeling, a high integration among the ASR components is indeed required to achieve better performance.

Strik (2001) mentioned that pronunciation variation modeling should be considered in three levels: the acoustic model, dictionary, and the language model. In the acoustic model, the pronunciation variation is accounted for by using triphone models and Gaussian mixer models (GMM) for the states' emission probabilities.

In the acoustic model, the triphones concept has been introduced to capture the phonological effects in continuous speech. So, instead of training a single HMM for each phoneme, several models are trained according to the context of the phoneme. That is, each model will be trained using one preceding and following phoneme context (Hirsimaki 2003). Hazen et al. (2005) examine the advantages and disadvantages of accounting for general phonological variation explicitly with phonological rules using distinct allophonic models versus implicitly within context-dependent models.

However, this book attempts to model Arabic phonological rules at two ASR levels: the dictionary and the language model. In fact, we need to measure the effect of phonological rules using the same acoustic model for a baseline and an enhanced system. Figure 4.3 shows the levels where we want to add the variants.

Most speech recognition systems rely on the pronunciation dictionaries that usually contain a few alternate pronunciations for most words. Additionally, the words' pronunciations in the dictionary are phonemically transcribed as if it will be uttered in isolation, which, consequently, leads to the cross-word problem. In fact, the utterance of a word in isolation is different from when the word is uttered in continuous speech. The cross-word problem occurs at word junctures and is represented by coarticulation of word boundary phonemes. Figure 4.4 shows the cross-word problem that occurs at the juncture between two adjacent words (w2 and w3). The merging between w2 and w3 forms a new phoneme sequence, which the recognizer cannot match to any single word in the pronunciation dictionary. Please notice that the Arabic text is read from right to left. However, we provide this example to be read as English from left to right for simplicity.

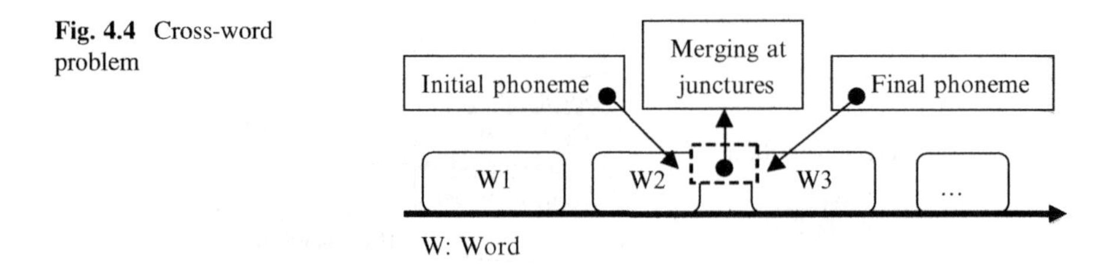

Fig. 4.4 Cross-word problem

Figure 4.4 shows that the continuous speech recognition systems face a discrimination problem when two consequent words are merged. Consequently, if the merged word is not available in the dictionary, errors may be presented in the recognition output. In addition to the cross-word problem, there are other problems that may occur in speech recognition systems.

With the successful use of context-dependent triphone to capture within-word and cross-word variations, the linguistic information can also be used for further enhancement for both variation types (i.e., cross-word and within-word). The language phonological rules could predict the variation at word's junctures. Consequently, knowing the potential variations may lead to having them correctly represented in the dictionary, language model, and/or acoustic model. Certainly, four well-known Arabic phonological rules can be applied: Idgham (merging), Iqlaab (changing), Hamzat Al-Wasl deleting, and merging of two unvoweled letters. Idgham, also called geminating or assimilation, is the merging of two consecutive letters. Iqlaab is the change of two consecutive letters into a completely different letter. Of course, in both Idgham and Iqlaab, the two letters belong to two separate words. Additionally, Idgham and Iqlaab also occur within word, as we described in Chap. 3. Hamzat Al-Wasl is an extra Hamza that helps to start pronouncing an unvoweled letter in continuous speech. Hamzat Al-Wasl can be omitted to merge the adjacent words. To avoid the problem of meeting two unvoweled (Saakin) letters, one of them can be omitted or voweled. Our focus will be on the modern standard Arabic (MSA), which is the recently spoken Arabic in news casts and formal speeches. MSA is learned in school, used in most newspapers, and is considered to be the official language in most Arabic-speaking countries (Lamel et al. 2009).

Initially, it is true, in principle, that pronunciation variation modeling enhances the ASR performance. Riley et al. (1998) considered multiple pronunciation variants as being the main problem of conversational ASR. Sloboda and Waibel (1996) demonstrated that having dictionaries, rich with more alternative pronunciations, is a key factor in improving the ASR performance. McAllester et al. (1998) showed that using pronunciation variations can produce much better performance than a baseline system which had no variants. Fosler-Lussier et al. (1999) showed that a mismatch between recognized phones and the phonemic transcription in the dictionary increases WER and deteriorates the performance. Saraçlar et al. (2000) showed that the ASR performance will be highly improved if there is a close match between the phonetic sequence recognized by the ASR decoder and the phonemic transcription in the dictionary. Amdal et al. (2000) indicated that an indirect data-driven approach used for modeling pronunciation variation for American English had reduced the WER. Kim et al. (2007) used an indirect data-driven approach for modeling pronunciation variation for Korean.

Riley and Ljolje (1995) demonstrated an indirect data-driven approach to train probabilistic, context-dependent phoneme-to-phone mappings to obtain a phonetic lexicon for native American speakers. Tajchman et al. (1995) provided an indirect data-driven approach to choose the useful pronunciations using probabilities associated with the generated rules. They also demonstrated the problem of blindly adding multiple pronunciations to a dictionary. Yang and Martens (2000) used a technique to get pronunciation variations by combining the rule-based data-driven method with the experts' knowledge for multi-pronunciation dictionary for different accents of nonnative Mandarin speakers. Seman and Kamaruzaman (2008) studied the within-word pronunciation variations in spontaneous Standard Malay. Lee and Chung (2007) implemented a rule-based pronunciation variants generator to produce a pronunciation lexicon with context-dependent multiple variants for Korean. Jurafsky et al. (2001) demonstrated three factors related to phonetic variation to study which of them might cause problems for a triphone-based recognizer. They investigated which kinds of pronunciation variation are well captured by triphone models, and which are not.

Even though our scope is cross-word variations, we briefly provide the literature of Arabic within-word variations followed then by the up-to-date cross-word variations literature. Alghamdi et al. (2009) developed an Arabic broadcast news transcription system. This transcription system has a phonetic dictionary which considers the variations for words that might be pronounced in different ways. Ali et al. (2009) presented a rule-based technique to generate Arabic phonetic dictionaries for a large vocabulary, speech recognition system. The system used classic Arabic pronunciation rules, common pronunciation rules of MSA, and morphologically driven rules. Al-Haj et al. (2009) provided a knowledge-based approach to add a number of pronunciation variations to the phonetic dictionary. They worked for Iraqi-Arabic speech and focused on short vowels. Biadsy et al. (2009) showed that the use of linguistic pronunciation rules can significantly improve phone recognition and word recognition results. They have developed a set of pronunciation rules that encapsulate some of MSA features for within-word variation. Billa et al. (2002) discussed a number of research issues for Arabic speech recognition, e.g., absence of short vowels in written text and the presence of compound words that are formed by the concatenation of certain conjunctions, prepositions, articles, and pronouns, as prefixes and suffixes to the word stem.

Phonological rules have been used to model the cross-word problem in many languages. For English, Weintraub et al. (1989) described performance improvements arising from detailed phonological modeling and from the incorporation of cross-word coarticulatory constraints. Giachin et al. (1991) demonstrated that the phonological rules are effective in providing corrective capability at low computational cost. Eleven phonological rules were implemented to handle coarticulation at word junctures. Beulen et al. (1998) presented an application of pronunciation variants in combination with phrases for a large vocabulary, continuous speech recognition system.

Their results showed that the improvement of the recognition of Wall Street Journal (WSJ) corpora was mainly due to the increased span of the language model. Ravishankar and Eskenazi (1997) demonstrated a method to automatically derive new word pronunciation, and context-dependent transformation rules for phone sequences. Their extracted rules can be applied even to words not in the training set, and across word boundaries, thus modeling context-dependent behavior. Nock and Young (1998) described a method for selecting a set of multiwords and for learning alternative pronunciations which are more representative of those found in fluent speech. Lestari and Furui (2010) focus on how to improve the performance of Indonesian ASR by alleviating the problem of pronunciation variation of proper nouns and foreign words (English words in particular). Uraga and Pineda (2002) proposed a method to generate multiple word pronunciations for Mexican Spanish. They used a set of rules for grapheme-to-phone conversion. Fosler-Lussier and Williams (1999) used decision-tree smoothing of phone recognition to combat confusability by augmenting a lexicon with variants using a confidence-based evaluation of potential variants. Tsai et al. (2007) proposed a three-stage framework for Mandarin Chinese to construct automatically the multiple-pronunciation dictionary while reducing the possible confusion caused.

Even though cross-word phenomenon has been widely investigated for English, research works have been found in other languages. AbuZeina et al. (2011) presented a method to model cross-word problem using a set of Arabic phonological rules. Kessens et al. (1999) investigated two approaches to model the cross-word problem for Dutch. In the first approach, cross-word processes were modeled by directly adding the cross-word variants to the lexicon, while the second approach was done by using multiwords which achieved better performance. Pousse and Perennou (1997) described two approaches that take into account pronunciation variants of words. They allowed handling (very common) phonological French phenomena like liaisons or mute-e elision. Boulianne et al. (2000) demonstrated finite-state transducers based approach to integrate automatic pronunciation rules and cross-word phenomena for French large vocabulary recognition systems.

In addition to the phonological rules, cross-word problem can be modeled using a data-driven approach. In this approach, no linguistic rules are used. It depends on training corpus transcription to merge words, such as merging small words. Sloboda and Waibel (1996) proposed to augment short words to create compound words for German speech recognition. They modeled the generated compound words at the dictionary level. Saon and Padmanabhan (2001) proposed a method to augment short words for US English. They modeled the data-driven variants on both the dictionary and the language model. Berton et al. (1996) introduced a method for handling the components of compounds rather than the compounds themselves. They proposed a method to decompose the OOV compound words to words already in the lexicon. Siegler and Stern (1995) applied a small number of data-driven rules to merge short words to produce compound words that added to the dictionary. Their method did not yield improvement in accuracy. Xin et al. (2009) investigated data-driven methods for expanding a Mandarin ASR lexicon with new multiple pronunciations and new compound words. The new compound words were replaced in the dictionary and represented in the language model. Finke and Waibel (1997) presented an approach to generate compound words using a set of data-driven rules. Examples of the extracted rules included G AH N AX as the alternative of GOING TO, and W AH N AX as the alternative of WANT TO. Zhang et al. (2000) investigated a statistical approach to extract Chinese compound words from very large corpora using mutual information and context dependency. Braga et al. (2001) established phonological rules that aim to solve the phonetic errors that appear in the beginning and end of words, namely, between connected words. They used a data-driven approach to capture the variations at word boundaries for Portuguese language.

4.2 Sources of Cross-Word Problem

The pronunciation dictionary is designed to be used with a particular set of words. However, an ASR decoder will not always be able to find a perfect match between the phonemic transcription in the dictionary and the phonetic transcription of a recognizer. This ambiguity increases the OOV, which is undesirable. OOV is a words' set of unsatisfied requests among all queries to the dictionary. In the case of unsatisfied request, another dictionary word with a nearest match pronunciation will be chosen, consequently increasing errors and reducing performance. Intuitively, to ameliorate the ASR performance, OOV should be reduced as much as possible. This reduction in OOV will alleviate the difficulties that may rise during the decoding process. OOV problem is partially solved by extending the dictionary with some good (i.e., high frequently occurred) variants. This technique is used in modern ASRs such as SPHINX which provide an option to add some variants for single words (i.e., within-word variations) such as

```
أدِنبرَة     E AE D IH M B R AA H (default)
أدِنبرَة(1)  E AE D IH M B R AA T
أدِنبرَة(2)  E AE D IH N B R AA H
أدِنبرَة(3)  E AE D IH N B R AA T
```

Cross-word variation occurs between two separated words to produce a new compound word that, of course, is not listed in the dictionary. For example, "مِرّفعِهَا" is a new merged word of "مِن رَفعِهَا", "عَمَّلاعِب" is a contraction of "عن ملاعب" and "مِمْبَين" is a coarticulation of "مِن بَين". In general, merging, contraction, coarticulation, and compounding are alternatives. There are four main sources of cross-word pronunciation variation problem: Idgham, Iqlaab, Hamzat Al-Wasl deletion, and merging of two unvoweled letters. Idgham has three types as shown in Fig. 4.5. Chapter 5 has more elaboration of these Arabic speech pronunciation variation phenomena.

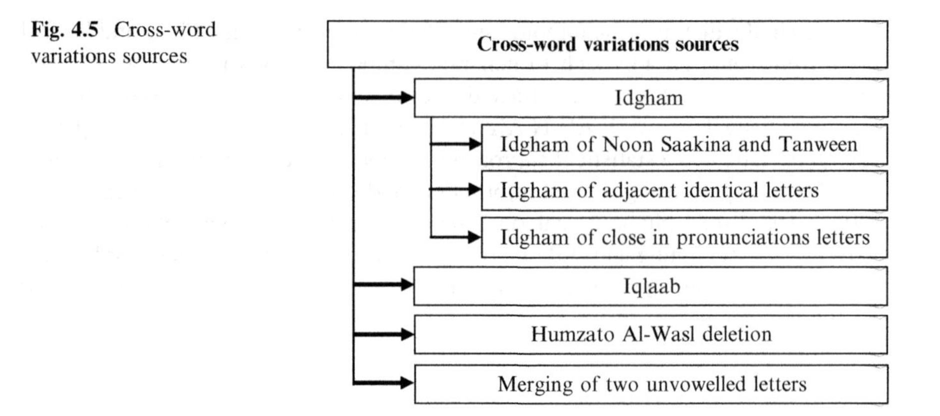

Fig. 4.5 Cross-word variations sources

4.3 Arabic Cross-Word Variations Examples

As described before, there are four main sources of cross-word variations. Here, we present some illustrative examples to show the effect of these variation sources. The explanation is performed with the help of the phoneme set described in Sect. 4.2. The examples aim to disclose the phonemes variations at the word junctures. Three illustrative cases will be presented: an Idgham case (Nuun Saakina or Tanween), an Iqlaab case, and an Idgham case (close-in-pronunciation letters case).

The actual speech pronunciation can be obtained using a phoneme recognizer. The phoneme recognizer output will then be compared with the canonical pronunciation to discover the occurred variations. So, a phoneme recognizer is used to produce the actual phoneme pronunciation, also called observation phonemes. Table 4.1 shows that the phoneme /N/ is converted to phoneme /AY/. This is an Idgham case where two letters are merging to generate a double letter of the second type (i.e., /AY/).

Table 4.2 shows that the phoneme /N/ is converted to /M/. This is an Iqlaab case in which two consequent letters are merged to generate a different letter.

Table 4.3 shows that the phoneme /T/ is converted to /D/. This is an Idgham case of two close-in-pronunciation letters.

Therefore, the one-to-one mapping that is usually used between the corpus transcription words and the dictionary entries cannot resolve the cross-word cases. As such, a technique for handling continuous speech cross-word merging is needed to achieve better performance. In Chap. 5, we introduce the Arabic phonological rules that were considered to model the cross-word phenomenon for Arabic speech.

Table 4.1 The effect of Idgham in Arabic speech

Rule name	Idgham (Nuun and Yaa)
Rule description	In Arabic, an unvoweled consonant N (نْ) at the end of a word can be merged with a voweled consonant Y (يَ) at the beginning of the next word to produce a new word with double consonant AY (يّ) at the connecting word's junctures
A speech signal with its transcription	وَأن يَحمِيَ المُستَهلِك. . . wa'n yaHmiya almustaHlik
Canonical pronunciation (dictionary)	W AE E AE **N Y**AE HH M IH Y AE E L M UH S T AE H L IH K
Actual pronunciation (phoneme recognizer)	W AE E **AY Y** AE HH M IH Y AE E L M UH S T AE H L IH K

Table 4.2 The effect of Iqlaab in Arabic speech

Rule name	Iqlaab
Rule description	In Arabic, an unvoweled consonant N (نْ) at the end of a word can be merged with a voweled consonant B (بَ) at the beginning of the next word to produce a new unvoweled consonant M(مْ) at the connecting word's junctures
A speech signal with its transcription	مِن بَينِهَا سِلتِل التَّابِعَة... min bayniha siltil altabi'a
Canonical pronunciation (dictionary)	M IH **N B** AY N IH H AE: S IH L T IH L E T AE: B IH AI AE H
Actual pronunciation (phoneme recognizer)	M IH **M B** AY N IH H AE: S AE L S TT R IX E L E AE T E AE B IH AI AE:

Table 4.3 Idgham of two close-in-pronunciation case

Rule name	Idgham two close-in-pronunciation letters (Taa and Dal)
Rule description	In Arabic, an unvoweled consonant at the end of a word Taa' (ت) can be merged with a close-in-pronunciation voweled consonant Daal (د) at the beginning of the next word to produce a double consonant of the second type
A speech signal with its transcription (a wav file)	أظهرت دراسة أعدها مجلس . . . aZharat dirasatun 'a'daha majlisu
Canonical pronunciation (dictionary)	E AE DH2 H AE R AA **T D** IH R AA: S AE T UH N E AE AI AE D AE H AE: M AE JH L IH S UH
Actual pronunciation (phoneme recognizer)	E AE DH2 UH H AE: R AA **D D** IH R AE SS AE TT UH E N E E AI D AE: H AE: M B AY E Z IH E S UH

References

AbuZeina D, Al-Khatib W et al (2011) Cross-word Arabic pronunciation variation modeling for speech recognition. Int J Speech Tech: 1–10

Alghamdi M, Elshafei M, Almuhtasib H (2009) Arabic broadcast news transcription system. Int J Speech Tech 10:183–195

Al-Haj H, Hsiao R, Lane I, Black A, Waibel A (2009) Pronunciation modeling for dialectal Arabic speech recognition, ASRU 2009: IEEE workshop, Italy

Ali M, Elshafei M, Alghamdi M, Almuhtaseb H, Alnajjar A (2009) Arabic phonetic dictionaries for speech recognition. J Inform Tech Res 2(4):67–80

Amdal I, Fosler-Lussier E (2003) Pronunciation variation modeling in automatic speech recognition. Telektronikk, 2.2003, pp 70–82

Amdal I, Korkmazskiy F, Surendran AC (2000) Joint pronunciation modeling of non-native speakers using data-driven methods, ICSLP, Beijing, China, pp 622–625

Benzeghiba M, De Mori R et al (2007) Automatic speech recognition and speech variability: a review. Speech Comm 49(10–11):763–786

Berton A, Fetter P, Regel-Brietzmann P (1996) Compound words in large-vocabulary German speech recognition systems. In: Proceedings of the fourth international conference on spoken language, 1996. ICSLP 96, vol 2. 3–6 Oct 1996, pp 1165–1168

Beulen K, Ortmanns S, Eiden A, Martin S, Welling L, Overmann J, Ney H (1998) Pronunciation modeling in the RWTH large vocabulary speech recognizer. In: Proceedings of the ESCA workshop modeling pronunciation variation for automatic speech recognition, pp 13–16

Biadsy F, Habash N, Hirschberg J (2009) Improving the Arabic pronunciation dictionary for phone and word recognition with linguistically-based pronunciation rules. The 2009 annual conference of the North American chapter of the ACL, Colorado, pp 397–405

Billa J, Noamany M et al (2002) Audio indexing of Arabic broadcast news. 2002 IEEE international conference on acoustics, speech, and signal processing (ICASSP)

Boulianne G, Brousseau J, Ouellet P, Dumouchel P (2000) French large vocabulary recognition with cross-word phonology transducers. ICASSP 3:1675–1678

Braga D, Freitas D, Barros MJ (2001) On the identification of word-boundaries using phonological rules for speech recognition and labeling phonological rules and trends in word endings. Forum American Bar Association, Chicago, IL

Finke M, Waibel A (1997) Speaking mode dependent pronunciation modeling in large vocabulary conversational speech recognition. In: Proceedings of EuroSpeech-97, Rhodes, pp 2379–2382

Fosler-Lussier E, Williams G (1999) Not just what, but also when: guided automatic pronunciation modeling for broadcast news. DARPA broadcast news workshop, Herndon, VA

Fosler-Lussier E, Greenberg S, Morgan N (1999) Incorporating contextual phonetics into automatic speech recognition. In: Proceedings of the international congress on phonetic sciences, pp 611–614

Giachin EP, Rosenberg AE et al (1991) Word juncture modeling using phonological rules for HMM-based continuous speech recognition. Comput Speech Lang 5(2):155–168

Hazen TJ, Hetherington IL, Shu H, Livescu K (2005) Pronunciation modeling using a finite-state transducer representation. Speech Comm 46(2):189–203

Helmer S (2001) Pronunciation adaptation at the lexical level. In: Proceedings ISCA ITRW workshop adaptation methods for speech recognition, Sophia Antipolis, France, 2001

Hirsimaki T (2003) A review: decision trees in speech recognition. Helsinki University of Technology, Finland

Hofmann H, Sakti S et al (2010) Improving spontaneous English ASR using a joint-sequence pronunciation model. 2010 4th international universal communication symposium (IUCS)

Jurafsky D, Ward W, Zhang J, Herold K, Yu X, Zhang S (2001) What kind of pronunciation variation is hard for triphones to model? In: Proceedings of ICASSP, 2001

Kessens JM, Wester M et al (1999) Improving the performance of a Dutch CSR by modeling within-word and cross-word pronunciation variation. Speech Comm 29(2–4):193–207

Kim M, Oh YR, Kim HK (2007) Non-native pronunciation variation modeling using an indirect data-driven method. In: Proceedings of ASRU, Japan, 2007

Lamel L, Messaoudi A et al (2009) Automatic speech-to-text transcription in Arabic. ACM Trans Asian Lang Inform Process 8(4):1–18

Lee K-N, Chung M (2007) Morpheme-based modeling of pronunciation variation for large vocabulary continuous speech recognition in Korean. IEICE Trans Inf Syst E90-D (7):1063–1072

Lestari D, Furui S (2010) Adaptation to pronunciation variations in Indonesian spoken query-based information retrieval. IEICE Trans Inf Syst E93.D(9):2388–2396

McAllester D, Gillick L, Scattone F, Newman M (1998) Fabricating conversational speech data with acoustic models: a program to examine model-data mismatch. In: Proceedings of ICSLP, Sydney, Australia, December 1998

Nock HJ, Young SJ (1998) Detecting and correcting poor pronunciations for multiword units. ESCA workshop, 1998

Pousse L, Perennou G (1997) Dealing with pronunciation variants at the language model level for automatic continuous speech recognition of French. In: Proceedings of Eurospeech-97, Rhodes, pp 2727–2730

Ravishankar M, Eskenazi M (1997) Automatic generation of context-dependent pronunciations. In: Proceedings of EuroSpeech-97, Rhodes, pp 2467–2470

Riley M, Ljolje A (1995) Automatic generation of detailed pronunciation lexicons. In: Lee CH, Soong FK, Paliwal KK (eds) Automatic speech and speaker recognition: advanced topics. Kluwer Academic, Boston

Riley M, Byrne W, Finke M, Khudanpur S, Ljolje A, McDonough J, Nock H, Saraclar M, Wooters C, Zavaliagkos G (1998) Stochastic pronunciation modelling from handlabelled phonetic corpora. In: Proceedings of ETRW on modeling pronunciation variation for automatic speech recognition, 1998, pp 109–116

Saon G, Padmanabhan M (2001) Data-driven approach to designing compound words for continuous speech recognition. IEEE Trans Speech Audio Process 9(4):327–332

Saraçlar M, Nock H, Khudanpur S (2000) Pronunciation modeling by sharing Gaussian densities across phonetic models. Comput Speech Lang 14:137–160
Seman N, Kamaruzaman J (2008) Acoustic pronunciation variations modeling for standard Malay speech recognition. Comput Inform Sci 1(4):112–120, ISSN 1913-8989
Siegler MA, Stern RM (1995) On the effects of speech rate in large vocabulary speech recognition systems. 1995 international conference on acoustics, speech, and signal processing, 1995. ICASSP-95
Sloboda T, Waibel A (1996) Dictionary learning for spontaneous speech recognition. In: Proceedings of ICSLP-96, Philadelphia, PA, USA, pp 2328–2331
Tajchman G, Foster E, Jurafsky D (1995) Building multiple pronunciation models for novel words using exploratory computational phonology. In EUROSPEECH-1995, pp 2247–2250
Tsai MY, Chou FC et al (2007) Pronunciation modeling with reduced confusion for Mandarin Chinese using a three-stage framework. IEEE Trans Audio Speech Lang Process 15(2): 661–675
Uraga E, Pineda LA (2002) Automatic generation of pronunciation lexicons for Spanish. In: Proceedings of the third international conference on computational linguistics and intelligent text processing, Springer, Netherlands, pp 330–338
Weintraub M, Murveit H et al (1989) Linguistic constraints in hidden Markov model based speech recognition. 1989 international conference on acoustics, speech, and signal processing, 1989. ICASSP-89
Wester M (2003) Pronunciation modeling for ASR—knowledge-based and data-derived methods. Comput Speech Lang 17:69–85
Wester M, Fosler-Lussier E (2000) A comparison of data-derived and knowledge-based modeling of pronunciation variation, ICSLP, Bejing, China, 2000
Xin L, Wen W et al (2009) Data-driven lexicon expansion for Mandarin broadcast news and conversation speech recognition. IEEE international conference on acoustics, speech and signal processing, 2009. ICASSP 2009
Yang Q, Martens J-P (2000) On the importance of exception and cross-word rules for the data-driven creation of Lexica for ASR. In: Proceedings of the 11th ProRisc workshop, 29 Nov–1 Dec, Veldhoven, The Netherlands, pp 589–593
Zhang J, Gao J et al (2000) Extraction of Chinese compound words: an experimental study on a very large corpus. In: Proceedings of the second workshop on Chinese language processing: held in conjunction with the 38th annual meeting of the association for computational linguistics, vol 12. Association for Computational Linguistics, Hong Kong, pp 132–139

Chapter 5
Modeling of Arabic Cross-Word Pronunciation Variations

This chapter discusses some Arabic phonological rules that can be used to capture the variations occurring at words' junctures. The rules include Idgham, Iqlaab, Hamzat Al-Wasl, and merging of two consecutive unvoweled letters. Then, an algorithm to model this problem is provided.

5.1 Introduction

Arabic speech recognition has gained increasing importance in the last few years. Speech recognition systems are often used as the front-end for many natural language processing (NLP) applications. These typical applications include voice dialing, call routing, data entry and dictation, information retrieval and extraction, command and control, computer-aided language learning, machine translation, etc. In fact, speech communication with computers is envisioned to be the dominant human–machine interface in the near future.

Arabic is a Semitic language spoken by more than 330 million people as a native language (Farghaly and Shaalan 2009). In this book, we focus on the modern standard Arabic (MSA) which is currently used in writing and most formal speech. MSA is the major medium of communication for public speaking and news broadcasting (Ryding 2005). The close relation between the Holy Qur'an and MSA phonological rules has helped to preserve MSA and its rules. The cross-word problem presented in the previous chapter will be modeled using MSA phonological rules. In the next sections, we explore selected MSA phonological rules and present the algorithm used to model the selected subset of MSA phonological rules. More information about MSA can be found in Ryding (2005).

D. AbuZeina and M. Elshafei, *Cross-Word Modeling for Arabic Speech Recognition*, SpringerBriefs in Electrical and Computer Engineering,
DOI 10.1007/978-1-4614-1213-7_5,

5.2 Arabic Phonological Rules

Arabic is a morphologically rich language in which many utterance changes can be captured by MSA phonological rules. The MSA phonological rules explained in this book include Idgham, Iqlaab, Hamzat Al-Wasl, and merging of two unvoweled letters. These four rules usually dominate the majority of cross-word variations that occur at the words' junctures.

In order to generate a compound word of two consecutive words, two letters are required: the final letter of the first word, and the initial letter of the second word. Modeling cross-word problem starts with the corpus transcription by searching for all cases that satisfy the modeled phonological rules. In Fig. 5.1, when words w3 and w4 satisfy the constraint of a particular phonological rule, such as Idgham or Iqlaab, the two words are merged.

The following subsections describe the MSA phonological rules that produce the cross-word problem.

5.2.1 *Idgham*

Idgham is a merging of two consecutive letters (could be in one word or in two separated words) to produce a single geminated letter. Idgham has three different forms: Idgham of Nuun Saakina and Tanween, Idgham of two consecutive identical letters, and Idgham of two letters close in pronunciation.

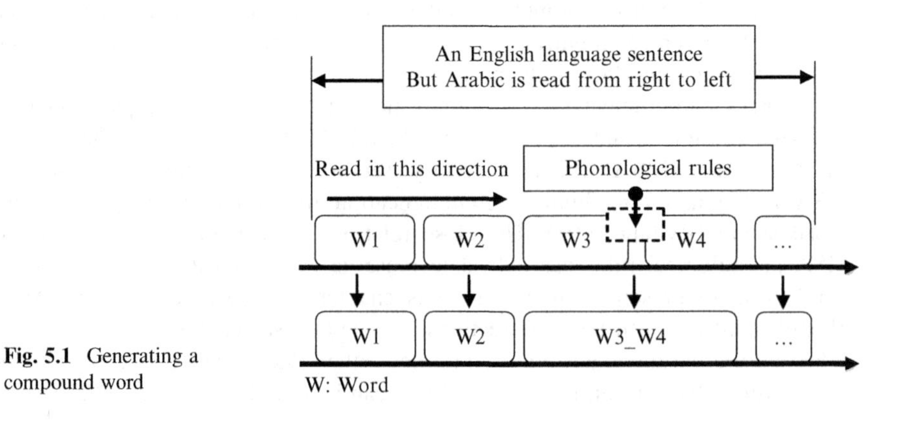

Fig. 5.1 Generating a compound word

Table 5.1 Idgham cases of Nuun Saakina

<table>
<tr><th>The final letter Of the first word (unvowelled)</th><th>Boundary</th><th>The initial letter Of the second word (Vowelled)</th></tr>
<tr><td>نْ / nuun</td><td>space</td><td>ي / yaa’</td></tr>
<tr><td colspan="3">وَمِن المُتَوَقَّع أنْ يَستَضِيفَ أكثَرَ مِن
wamina almutawaq‘ an yastaDiifa ’kthar min
وَمِن المُتَوَقَّع أيَّستَضِيفَ أكثَرَ مِن
wamina almutawaq‘ ayyastaDiifa ’kthar min</td></tr>
<tr><td>نْ / nuun</td><td>space</td><td>ر / raa’</td></tr>
<tr><td colspan="3">بَعدَ شَهرٍ وَاحِدٍ مِنْ رَفعِهَا لِلحَظر
b‘d shahrin waHidin min raf‘iha lilhazr
بَعدَ شَهرٍ وَاحِدٍ مِرَّفعِهَا لِلحَظر
b‘d shahrin waHidin min raf‘iha lilhazr</td></tr>
<tr><td>نْ / nuun</td><td>space</td><td>م /miim</td></tr>
<tr><td colspan="3">تُجبِرُهَا عَلَى الإبتَعَادِ عَنْ مَلاعِبِ التِّنِس
tujbiruha ‘ala al’bti‘adi ‘an mula‘ibi altanis
تُجبِرُهَا عَلَى الإبتَعَادِ عَمَّلاعِبِ التِّنِس
tujbiruha ‘ala al’bti‘adi ‘ammula‘ibi altanis</td></tr>
<tr><td>نْ / nuun</td><td>space</td><td>ل / laam</td></tr>
<tr><td colspan="3">مُؤكِّدًا إستِعَادَتَهُ بَعضًا مِنْ لَيَاقَتِهِ البَدَنِيَّة
mu’kidan ’sti ‘adatahu b‘dan min layaqatihi ’lbadaniya
مُؤكِّدًا إستِعَادَتَهُ بَعضًا مِلَّيَاقَتِهِ البَدَنِيَّة
mu’kidan ’sti ‘adatahu b‘dan milayaqatihi ’lbadaniya</td></tr>
<tr><td>نْ / nuun</td><td>space</td><td>و / waaw</td></tr>
<tr><td colspan="3">أكثَرَ مِنْ وَاحِد وَسِتِينَ مِليُونَ شَخص
akthara min waHid wasitiin milyon shakhS
أكثَرَ مِوَّاحِدٍ وَسِتِينَ مِليُونَ شَخص
akthara miwwaHid wasitiin milyon shakhS</td></tr>
<tr><td>نْ / nuun</td><td>space</td><td>ن / nuun</td></tr>
<tr><td colspan="3">مَنعَ الجَمَاهِير مِنْ نُزُولِ أرضِ المَلعَب
man ‘a aljamahiir min nuzwl ’rd almal‘ab
مَنعَ الجَمَاهِير مِنُّزُولِ أرضِ المَلعَب
man ‘a aljamahiir minnuzwl ’rd almal‘ab</td></tr>
</table>

5.2.1.1 Idgham of Nuun Saakina and Tanween

It is a merging between unvoweled nuun (Nuun Saakina: نْ) or Tanween (ٌ ، ً ، ٍ) and one of the following voweled letters (ي، ر، م، ل، و، ن). Table 5.1 shows examples of unvoweled nuun followed by the letters of Idgham {(ي، ر، م، ل، و، ن)}. For each case in Table 5.1, the first sentence is the original sentence as it is in the corpus transcription, and the second one is the sentence after the merging process. Table 5.1 provides examples of Nuun Saakina. Tanween (ٌ ، ً ، ٍ) is similar.

5.2.1.2 Idgham of Two Consecutive Identical Letters (Idgham almutmathlan <> إدغام المتماثلان)

It is a merging between two consecutive identical letters shown in the following list {{ب, ت, ث, ج, ح, خ, د, ذ, ر, ز, س, ش, ص, ض, ط, ظ, ع, غ, ف, ق, ك, ل, ن}}. The rule means that any unvoweled Arabic letter followed by the same Arabic voweled letter will be doubled in a single merged word. Note that {{ ا, و, ي }} are not included in the list. Table 5.2 shows merging cases of consecutive identical letters.

Table 5.2 Idgham of two consecutive identical letters

The final letter Of the first word (Unvowelled)	Boundary	The initial letter Of the second word (Vowelled)
سْ / Siin	space	س / Siin
إنَّ هَذَا المَجلِسْ سَيُشرِفُ عَلَى الثَّروَةِ النِّفطِيَّة 'na hadha 'lmajlis sayushrifu 'ala 'ltharwati 'lnifTiya إنَّ هَذَا المَجلِسَّيُشرِفُ عَلَى الثَّروَةِ النِّفطِيَّة 'na hadha 'lmajlissayushrifu 'ala 'ltharwati 'lnifTiya		
عْ / 'ayn	space	ع / 'ayn
خَاصَّةً مَعْ عَدَم تَوَفُّرِ أمَاكِنِ لِلبِنَاءِ khaSatan ma' 'adam tawafur 'makin lilbna' خَاصَّةً مَعَّدَم تَوَفُّرِ أمَاكِنِ لِلبِنَاءِ khaSatan ma''adam tawafur 'makin lilbna'		
لْ / laam	space	ل / laam
التَّقرِيرِ التَّالِي لِلزَّمِيلْ لُطفِي المَسعُودِي 'ltaqriir 'ltaly lilzamyl lutfy almas'wdy التَّقرِيرِ التَّالِي لِلزَّمِيلُّطفِي المَسعُودِي 'ltaqriir 'ltaly lilzamyllutfy almas'wdy		
تْ / taa'	space	ت / taa'
وَبَلَغَتْ تَكلِفَةُ إستِحوَاذِ شَرِكَةِ المَملَكَةِ wabalaghat taklifatu 'stihwadhi sharikati 'lmamlakati وَبَلَغَتَّكلِفَةُ إستِحوَاذِ شَرِكَةِ المَملَكَةِ wabalaghattaklifatu 'stihwadhi sharikati 'lmamlakati		
فْ / Faa'	space	ف / Faa'
المُتَوَقَّع لِلوَظَائِفْ فِي الاِقتِصَادِ الأمرِيكِيِّ 'lmutawaqa' lilwaZa'f fy 'l'iqtiSadi 'l'mryky المُتَوَقَّع لِلوَظَائِفِّي الاِقتِصَادِ الأمرِيكِيِّ 'lmutawaqa' lilwaZa'ffy 'l'iqtiSadi 'l'mryky		

5.2.1.3 Idgham of Two Close-in-Pronunciation Letters (Idgham almutajanisan <>إدغام المتجانسان)

It is a merging between two consecutive different letters that are close in pronunciation. Table 5.3 shows these rules with examples.

5.2.2 Iqlaab

Iqlaab is a replacement of Nuun Saakinah (نْ) or Tanween that comes before voweled Baa (بـ) by Meem Saakinah (مْ). The following are examples of Iqlaab. Note that instead of geminating the connecting letter, it is unvoweled (مْ).

لِلاشتِرَاكِ فِي المَزَادِ العَالَمِيِّ مِنْ بَيْنِ سَبعَةِ

lil'shtiraki fy 'lmazadi 'l'alamyi min bayni sab'ati

لِلاشتِرَاكِ فِي المَزَادِ العَالَمِيِّ مِمْبَيْنِ سَبعَةِ

lil'shtiraki fy 'lmazadi 'l'alamyi mimbayni sab'ati

الجَولَةِ الثَّانِيةِ مِنْ بُطُولَةِ العَالَم

'ljawlati 'lthaniya min buTwlati 'l'alam

الجَولَةِ الثَّانِيةِ مِمْبُطُولَةِ العَالَم

'ljawlati 'lthaniya mimbuTwlati 'l'alam

أهلاً بِكُم إِلَى النَّشرَة الاِقتِصَادِيَّة

'hlan bikum 'la 'lnashrat 'l'qtiSadiya

أهلامْبِكُم إِلَى النَّشرَة الاِقتِصَادِيَّة

'hlambikum 'la 'lnashrat 'l'qtiSadiya

5.2.3 Hamzat Al-Wasl Deletion

Hamzat Al-Wasl is an extra Hamza that is used in continuous speech. It is pronounced when starting, and ignored when continuing. Hamzat Al-Wasl is used with names, verbs, and participles.

Table 5.3 Idgham of two close-in-pronunciation letters

Rule	Initial letter Of first word (Unvowelled)	Final letter Of second word (Vowelled)	Connecting letter (Double)
1	taa'/ تْ	daal / د	daal / دّ
	كَشَفَتْ دِرَاسَةٌ حَدِيثَة أنَّ بِرِيطَانيَا kashafat dirasatun Hadythatun 'na brytanya كَشَفَدَّرَاسَةٌ حَدِيثَة أنَّ بِرِيطَانيَا kashafaddirasatun Hadythatun 'na brytanya		
2	taa' / تْ	Taa' /ط	Taa / طّ
	تَعتَزِمُ شَرِكَةُ طَيَرَانِ الإمَارَاتْ طَلَبَ t'tazim sharikatu Tayaran 'l'marat Talab تَعتَزِمُ شَرِكَةُ طَيَرَانِ الإمَارَاطَّلَبَ t'tazim sharikatu Tayaran 'l'maraTalab		
3	daal / دْ	taa' /ت	taa' / تّ
	يَقُولُ مُنتَقِدُوهَا إنَّهَا قَدْ تُؤَجِّجُ التَّضَخُم yaqwlu muntaqidwha 'naha qad tu'jiju 'ltaDakhum يَقُولُ مُنتَقِدُوهَا إنَّهَا قَتُّؤَجِّجُ التَّضَخُم yaqwlu muntaqidwha 'naha qattu'jiju 'ltaDakhum		
4	baa' / بْ	miim / م	miim / مّ
	يَا بُنَيَّ ارْكَبْ مَعَنَا ya bunya 'rkab ma'na يَا بُنَيَّ ارْكَمَّعَنَا ya bunya 'rkamma' na		
5	dhaal / ذْ	Zaa /ظ	Zaa / ظّ
	وَلَوْ أنَّهُمْ إِذْ ظَلَمُوا أنْفُسَهُمْ walaw 'nahum 'Z Zalamw 'nfusahm وَلَوْ أنَّهُمْ إِظَّلَمُوا أنْفُسَهُمْ walaw 'nahum 'ZZalamw 'nfusahm		
6	qaaf / قْ	kaaf /ك	kaaf / كّ
	أعلَنَ وَزِيرُ الاِتِّصَالات المِصرِيّ طَارِقْ كَمَال ''lana wazyru 'l'tiSalat 'lmaSry Tariq kamal أعلَنَ وَزِيرُ الاِتِّصَالات المِصرِيّ طَارِكَّمَال عَن طَرح ''lana wazyru 'l'tiSalat 'lmaSry Tarikkamal		
7	thaa' / ثْ	dhaal /ذ	dhaal / ذّ
	أوْ تَتْرُكْهُ يَلْهَثْ ذَلِكَ مَثَلُ الْقَوْم 'w tatrukhu yalhath dhalk mathalu 'lqawm أوْ تَتْرُكْهُ يَلْهَذَّلِكَ مَثَلُ الْقَوْم 'w tatrukhu yalhatdhdhalk mathalu 'lqawm		
8	laam /ل	raa' / ر	raa' / ر
	التَقرِير لِلزُمِيلْ رَامِي إبرَاهِيم 'ltaqryr llzamyl ramy 'brahym التَقرِير لِلزُمِيرَّامِي إبرَاهِيم 'ltaqryr llzamyrramy 'brahym		

Table 5.4 The cases of Hamzat Al-Wasl in Arabic

Verb	Three letter root	Four letter root	Five letter root	Six letter root
Past tense	No	No	Yes افْتَرَى اعتَدَى	Yes اسْتَكْبَرَ استَسقَى
Present tense	No	No	No	No
Command	Yes ابْنِ اذهَب	No	Yes انطَلِقُوا انتَهُوا	Yes اسْتَغْفِرْ استَئجِرهُ

Names: there are nine cases where Hamzat Al-Wasl is found in Arabic nouns. These names are ابن ،ابنت ، اثنتين ، اثنين ، اسم، امرأت ،امرؤ.. Additionally, it includes the original noun (المصدر) of a five-letter past tense verb (e.g., افتراء), and the original noun of a six-letter past tense verb (استغفار). The following examples show the existence of Hamzat Al-Wasl and how it could be ignored to generate a compound word.

أربَعُونَ شَرِكَةً تُمَثِّل اثنَتَينِ وَعِشرِينَ دَولَة
'rba'wna sharikatan tumathil 'thnatayni wa'ishriin dawla
أربَعُونَ شَرِكَةً تُمَثِّلثنَتَينِ وَعِشرِينَ دَولَة
'rba'wna sharikatan tumathilthnatayni wa'ishriin dawla

فِي بِنَاءِ مُجَمَّع فَخم فِي ضَاحِيَةِ خَلِيج البَحرَين تَحتَ اسم
fy bna' mujama'in fakhm fy Dahiyati khaliij 'lbahryn tahta 'sm
فِي بِنَاءِ مُجَمَّع فَخم فِي ضَاحِيَةِ خَلِيج البَحرَين تَحتَسم
fy bna' mujama'in fakhm fy Dahiyati khaliij 'lbahryn tahtasm

بِمَا نِسبَتُهُ تِسعَةً فَاصِل اثنَين وَعِشرِين فِي المِئَة
bma nisbatuhu tis'atn faSil 'thnyn wa'ishryn fy 'lmadiina
بِمَا نِسبَتُهُ تِسعَةً فَاصِلثنَين وَعِشرِين فِي المِئَة
bma nisbatuhu tis'atn faSilthnyn wa'ishryn fy 'lmadiina

Verbs: while Hamzat Al-Wasl does not occur in Arabic present tense verbs, it occurs in some past and command tense verbs. However, it does not occur in three- and four-letter root in the past tense. Additionally, it also does not occur in four-letter root in command verbs. Table 5.4 shows Hamzat Al-Wasl occurrences in different kinds of verbs with examples. "No" in Table 5.4 means that no occurrence of Hamzat Al-Wasl in that kind of verbs.

The following are four examples of Hamzat Al-Wasl in Arabic sentences. Note that Hamzat Al-Wasl has been omitted in the merging case.

رَبِّ ابْنِ لِي عِنْدَكَ بَيْتًا فِي الْجَنَّةِ

rabi 'bn ly ‘indaka baytan fy 'ljanati

رَبِّبْنِ لِي عِنْدَكَ بَيْتًا فِي الْجَنَّةِ

rabibn ly ‘indaka baytan fy 'ljanati

قَالَتْ إِحْدَاهُمَا يَا أَبَتِ اسْتَأْجِرْهُ

qalat 'Hdahuma ya 'bati 'sta'jirhu

قَالَتْ إِحْدَاهُمَا يَا أَبَتِسْتَأْجِرْهُ

qalat 'Hdahuma ya 'batista'jirhu

وَجَحَدُوا بِهَا وَاسْتَيْقَنَتْهَا أَنفُسُهُمْ

wajaHadw bha wa'stayqanatha anfusuhm

وَجَحَدُوا بِهَا وَسْتَيْقَنَتْهَا أَنفُسُهُمْ

wajaHadw bha wastayqanatha anfusuhm

وَإِذِ اسْتَسْقَى مُوسَى لِقَوْمِهِ فَقُلْنَا اضْرِبْ

wa'dh 'stasqa mwsa liqawmihi faqulna 'drib

وَإِذِسْتَسْقَى مُوسَى لِقَوْمِهِ فَقُلْنَا اضْرِبْ

wa'dhstasqa mwsa liqawmihi faqulna 'drib

Participles: Hamzat Al-Wasl occurs exclusively in determiner (الـ التعريف). For example,

بَينَهُم خَمسَةَ عَشَرَ وَزِيرًا ضِدَّ القَانُون

baynahum khamsata ‘ashara waziiran Did 'lqanwn

بَينَهُم خَمسَةَ عَشَرَ وَزِيرًا ضِدَّلقَانُون

baynahum khamsata ‘ashara waziiran Didlqanwn

5.2.4 Merging of Two Unvoweled Letters

Arabic continuous speech has no case where two consecutive unvoweled letters are to be pronounced except at the end of utterance. Therefore, two options are available to avoid the case of meeting two unvoweled letters: first, by deleting the first unvoweled letter such as

وَاسْتَبَقَا الْبَابَ وَقَدَّتْ قَمِيصَهُ مِن دُبُرٍ

wa'stabaqa 'lbab waqadat qamiiSahu min duburi

وَاسْتَبَقَلْبَابَ وَقَدَّتْ قَمِيصَهُ مِن دُبُرٍ

wa'stabaqalbab waqadat qamiiSahu min duburi

In the above example, note that the first unvoweled letter is the last letter in the word (وَاشْتَبَقَا) and the second unvoweled letter is (لْ) in the word (الْبَابَ). Hamzat Al-Wasl which is the first letter in the word (الْبَابَ) is an extra Hamza that has no effect in continuous speech as we mentioned in the previous section.

Second, by having the first letter to be voweled such as

قُلِ ٱدْعُوا ٱللَّهَ أَوِ ٱدْعُوا ٱلرَّحْمَٰنَ
qul 'd'w 'llaha 'w 'd'w 'lraHman
قُلِدْعُوا ٱللَّهَ أَوِدْعُوا ٱلرَّحْمَٰنَ
quld'w 'llaha 'wd'w 'lraHman

Note that the first unvoweled letter is voweled using Kasra and then merged with the second unvoweled letter (دْ) in the next word (ادْعُوا). In fact, Kasra, Damma, and Fatha can be used. The following example uses Damma:

ثُمَّ رَدَدْنَا لَكُمُ الْكَرَّةَ عَلَيْهِمْ
thuma radadna lakum alkarata 'alyhim
ثُمَّ رَدَدْنَا لَكُمُلْكَرَّةَ عَلَيْهِمْ
thuma radadna lakumlkarata 'alyhim

Heloz (2008) provide four rules to specify when to use Kasra, Damma, and Fatha to avoid meeting of two unvoweled letters in Arabic.

5.3 Modeling Arabic Cross-word Variation

Two ASR knowledge sources will be used in our method: the dictionary and the language model. The dictionary will be expanded by adding all cross-word candidates collected from the corpus transcription. The language model will also be expanded according to the cross-word cases found in the corpus transcription. The following are the steps required in our method:

Step 1: Have the corpus transcription ready to distil the cross-word candidates. Figure 5.2 shows a part of our corpus transcription. In Fig. 5.2, we chose small sentences for illustration purpose.

Step 2: Specify the phonological rules that you would like to apply. As we saw in the previous section, there are many rules that can be applied. In Chap. 6, we demonstrated the results of testing only Idgham and Iqlaab.

Step 3: Develop a program with a suitable programming language to extract the compound words using the selected rules.

Step 4: The identified cases of compound words are then added to the corpus transcription within their sentences. Figure 5.3 shows some sentences which include compound words. Note that the original sentences (i.e., without merging) remain in the enhanced corpus transcription.

. . .

بَعدَ أن أعلَنَت الشَّركَة الأورُوبيَّة

وَتَحُثُّهُمَا عَلَى أن يُفَكِّرَا فِي خَطِّ أنَابِيبِ غَازٍ بَدِيل

الَّتِي شَهِدَت زِيَادَةَ عَدَدِ الوَحَدَاتِ المَبِيعَة

كَمَا تَمَّ الإعلانُ عَن وُصُول إجمَالِيِّ عَدَدِ أعضَاءِ المَجلِسِ العَام لِلبُنُوك

وَكَانَت اليَابَان قَد جَدَّدَت يَوم الجُمعَة

وَذَلِكَ بَعدَ أسبُوع مِن التَّفجِيرَينِ اللَّذَينِ قَطَعَا خَطَّ الأنَابِيب

رَأيُ نَقَابَةِ الطَّيَّارِينَ لَم يَأتِ مُتَطَابِقًا مَع رَأي إدَارَةِ المَطَار

. . .

Fig. 5.2 A sample of the transcription corpus used

. . .

كَشَفَتْ دِرَاسَةٌ حَدِيثَة أنَّ بِرِيطَانيَا

كَشَفَدِّرَاسَةٌ حَدِيثَة أنَّ بِرِيطَانيَا

تَعتَزِمُ شَرِكَةُ طَيَرَانِ الإمَارَاتْ طَلَبَ

تَعتَزِمُ شَرِكَةُ طَيَرَانِ الإمَارَاطَّلَبَ

أعلَنَ وَزِيرُ الاِتِّصَالات المِصرِيِّ طَارِقْ كَمَال عَن طَرحِ

أعلَنَ وَزِيرُ الاِتِّصَالات المِصرِيِّ طَارِكَّمَال عَن طَرحِ

. . .

Fig. 5.3 A sample of the enhanced corpus transcription

Partial Phonetic Dictionary

. . .

مِرَّفعِهَا M IH R AA F AI IH H AE:

عَمَّلاعِبِ AI AE M AE L AI IH B IH

مِمْبَينِ M IH M B AE AY N IH

. . .

Fig. 5.4 A sample of the dictionary entries

Step 5: We use the enhanced corpus transcription generated in Step 4 to build the enhanced dictionary. Figure 5.4 shows some entries of the enhanced dictionary. The figure shows some cross-word entries, even though it contains all words of the enhanced corpus transcription (i.e., merged and non-merged words).

Step 6: Build the language model according to the enhanced corpus transcription. This means that the compound words in the enhanced corpus transcription will be involved in the unigrams, bigrams, and trigrams of the language model.

Step 7: Once the enhanced dictionary and the enhanced language model are ready, the recognition process (decoding) can be started.

Step 8: After recognition process, the recognition result is scanned for decomposing compound words to their original state (two separated words). This process is done using a lookup table such as

مِرَّفعِهَا (mirraf'iha) → مِنْ رَفعِهَا (min raf'iha)
عَمَّلاعِبِ ('ammala'ib) → عَنْ مَلاعِبِ ('an mala'ib)
مِمْبَيْنِ (mimbayn) → مِنْ بَينِ (min bayn)

It is worth noting that each transformation case is represented in a separate sentence. For example, the following sentence

سَتَصرِفُ خِلالَ أيَّامِ رَاتِبَ شَهرٍ وَاحِدٍ لِرُبعِ مُوَظَّفِيّ
satasrifu khilala 'yamin ratiba shahrin waHidin lirub'i muwaZafyi

has been modeled using four separated sentences (the original one plus 3 transformation cases), as shown below.

1) سَتَصرِفُ خِلالَ أيَّامِ رَاتِبَ شَهرٍ وَاحِدٍ لِرُبعِ مُوَظَّفِيّ
satasrifu khilala 'yamin ratiba shahrin waHidin lirub'i muwaZafyi

2) سَتَصرِفُ خِلالَ **أيَّامرَّاتِبَ** شَهرٍ وَاحِدٍ لِرُبعِ مُوَظَّفِيّ
satasrifu khilala **'yamratiba** shahrin waHidin lirub'i muwaZafyi

3) سَتَصرِفُ خِلالَ أيَّامِ رَاتِبَ شَهرٍ **وَاحِدلِّرُبعِ** مُوَظَّفِيّ
satasrifu khilala 'yamin ratiba shahrin **waHidlirub'i** muwaZafyi

4) سَتَصرِفُ خِلالَ أيَّامِ رَاتِبَ **شَهروَّاحِدٍ** لِرُبعِ مُوَظَّفِيّ
satasrifu khilala 'yamin ratiba **shahrwaHidin** lirub'i muwaZafyi

The steps of modeling cross-word variations are described in the following algorithm:

Cross-word pronunciation variation algorithm
w1 = first word in the transcription
Repeat
  w2 = next word in the transcription
  If w1&w2 satisfy one of the applied rules
    Generate the new compound word w1–w2
    Add w1–w2 to the Baseline dictionary
    Represent the w1–w2 in the transcription
    w1 = next word in the transcription
  Else
    w1 = w2
Until there are no more words in the transcription
Build the language model based on new transcription

The flowchart of the cross-word pronunciation variation is described in Fig. 5.5.

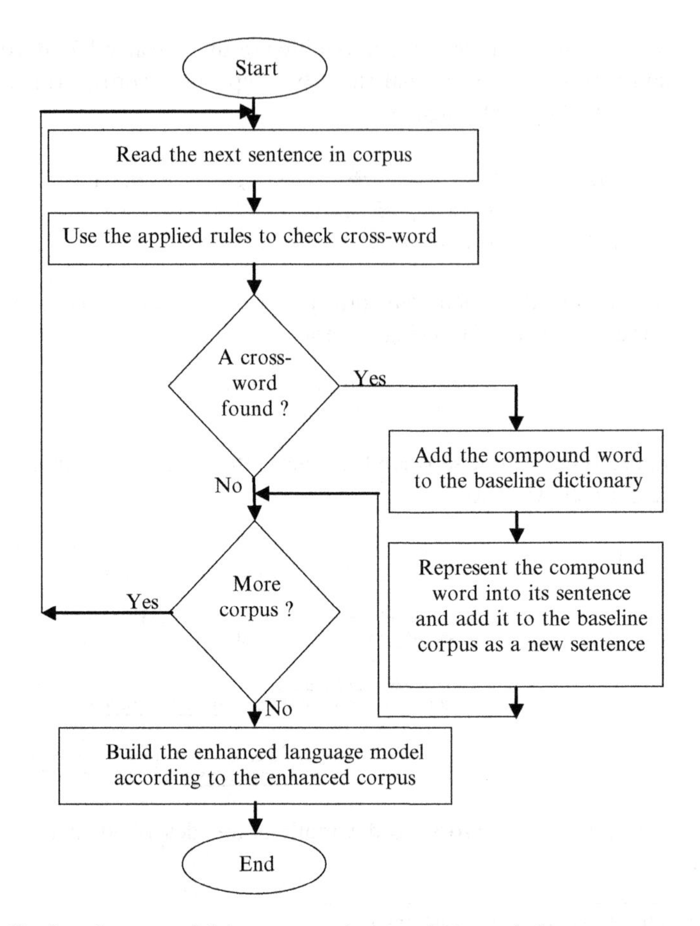

Fig. 5.5 The flowchart to model the cross-word pronunciation variation

References

Farghaly A, Shaalan K (2009) Arabic natural language processing: challenges and solutions 8(4): 1–22

Heloz A (2008) Almoyassar Almofeed fe Ilm Altajweed, Jordan, http://www.islamhouse.com/p/320902

Ryding KC (2005) A reference grammar of modern standard arabic (reference grammars). Cambridge University Press, Cambridge

Chapter 6
Performance and Evaluation

This chapter presents the results achieved by modeling cross-word pronunciation variation problem of MSA. We practically investigated two MSA phonological rules (Idgham and Iqlaab) which significantly enhanced recognition accuracy. Three ASR's metrics were measured: word error rate (WER), out of vocabulary (OOV), and perplexity (PP).

6.1 Experiment Setup

The baseline system described in Chap. 3 was used to measure the enhancement achieved using MAS phonological rules. Even though we presented four rules, we experimentally checked the effect of two of them: Idgham and Iqlaab rules (i.e., we have not tested Hamzat Al-Wasl and merging of two unvoweled letters rules). To measure the enhancement, we split the audio recordings into training set and testing set. The training set contained around 4.3 h of audio, while the testing set contained the remaining 1.1 h. We used the CMU language toolkit to build a statistical language model from the transcription of the full diacritized transcription of 5.4 h of the audio. For more information about the sphinx-III parameters used in the baseline, please refer to Sect. 1.2.1, the front-end signal processing. The training stage shows that the number of triphones came to 10,164 as displayed in Table 6.1.

D. AbuZeina and M. Elshafei, *Cross-Word Modeling for Arabic Speech Recognition*, SpringerBriefs in Electrical and Computer Engineering,
DOI 10.1007/978-1-4614-1213-7_6,

Table 6.1 Number of triphones for each phoneme in the baseline system

Phoneme	Triphones	Phoneme	Triphones
/AE/	542	/DH/	65
/AE:/	389	/R/	460
/AA/	96	/Z/	192
/AH/	64	/S/	302
/UH/	487	/SH/	144
/UW/	257	/SS/	156
/UX/	70	/DD/	137
/IH/	657	/TT/	161
/IY/	372	/DH2/	41
/IX/	85	/AI/	289
/AW/	77	/GH/	83
/AY/	104	/F/	286
/E/	479	/Q/	238
/B/	324	/K/	225
/T/	393	/L/	560
/TH/	106	/M/	344
/JH/	181	/N/	454
/HH/	195	/H/	258
/KH/	130	/W/	187
/D/	356	/Y/	218

6.2 Performance Metrics

Three performance metrics were used to measure the performance enhancement: the word error rate (WER), out of vocabulary (OOV), and perplexity (PP).

6.2.1 *Word Error Rate*

WER is a common metric to measure the performance of ASRs. WER is computed using the following formula:

$$\mathrm{WER} = \frac{(S + D + I)}{N},$$

where

- S is the number of substitution errors,
- D is the number of the deletion errors,
- I is the number of the insertion errors,
- N is the number of words in the reference.

The word accuracy can also be measured using WER as the following formula:

$$\text{Word accuracy} = 1 - \text{WER}.$$

We developed a tool which compares the recognition output and the reference text for Arabic text. The tool can be set to compare fully vocalized text or the non-vocalized text. The tool compares the two texts line by line and computes the number of substitution errors (*S*), deletion errors (*D*), and insertion errors (*I*).

6.2.2 Out of Vocabulary

OOV is a metric to measure the performance of ASRs. OOV is known as a source of recognition errors, which in turn could lead to additional errors in the words that follow (Gallwitz et al. 1996). Hence fore, increasing OOVs plays a significant role in increasing WER and deteriorating performance. Our ASR system is based on a closed vocabulary, so we assume that there are no unknown words. The closed vocabulary assumes that all words of the testing set are already included in the dictionary. Jurafsky and Martin (2009) explore the differences between open and closed vocabulary. In our method, we calculate OOV as the percentage of recognized words that do not belong to the testing set, but to the training set.

$$\text{OOV} = \frac{\text{None testing set words}}{\text{Total words in the testing set}}.$$

6.2.3 Perplexity

The perplexity of the language model is defined in terms of the inverse of the average log likelihood per word (Jelinek 1999). It is an indication of the average number of words that can follow a given word, a measure of the predictive power of the language model (Saon and Padmanabhan 2001). Measuring the perplexity is the common way to evaluate *n*-gram language model. It is a way to measure the quality of a model independent of any ASR system. The measurement is performed on the testing set. The lower perplexity system is considered better than one of higher perplexity. The perplexity formula is

$$\text{PP}(W) = \sqrt[N]{\frac{1}{P(w1, w2, \ldots, wN)}},$$

where PP is the perplexity, *P* is the probability of the word set to be tested $W = \text{w1}$, w2, ..., w*N*, and *N* is the total number of words in the testing set.

6.3 Evaluation and Results

The metrics (WER, OOV, and perplexity) explained in the previous section were measured. The enhanced system achieved a WER of 9.91% on the testing set. The WER decreased by 2.3% compared to the WER of the baseline system which was 12.21%, as summarized in Table 6.2.

The OOV was also measured for both systems. It was found that the baseline system has an OOV equal to 3.53%. The OOV was then reduced to 2.89% in the enhanced system. The OOV of both the systems (baseline and enhanced) was measured by dividing none testing set words over the total words in the testing set as follows:

$$\text{OOV (baseline system)} = \frac{\text{None testing set words}}{\text{Total words in the testing set}} = \frac{328}{9{,}288} \times 100 = 3.53\%,$$

$$\text{OOV(enhanced system)} = \frac{\text{None testing set words}}{\text{Total words in the testing set}} = \frac{269}{9{,}288} \times 100 = 2.89\%.$$

Clearly, the enhanced system is better.

Regarding perplexity, it was measured for both systems (baseline and enhanced) and found to be 34.08 and 4.00, respectively. The measurement was performed on the testing set which contains 9,288 words. So, the enhanced system is clearly better as the lower perplexity is better. The reason why both perplexities are low is that the specific domains of our corpus are limited to the economics and sports news. For more information about our corpus, please refer to Sect. 3.1.

The three metrics used to measure the performance shows the enhancement occurred using the applied phonological rules (i.e., Idgham and Iqlaab). To check whether the achieved enhancement is significant, we used the performance detection method suggested by Plötz (2005). We used a level of confidence of 95% to specify the confidence intervals. We also used the total number of words in the testing set and the word accuracy of the baseline system (87.79%). We found the boundaries of the confidence interval to be (+/−) 0.67%. Since the enhanced system achieved a word accuracy of 90.09%, we can conclude that our enhancement shows statistically significant improvement as the achieved enhancement (2.3%) is more than 0.67%.

Table 6.2 Performance improvement using phonological rules

Dictionary used	WER %
Baseline	12.21
Enhanced	9.91
Enhancement → 2.30	

6.4 Performance Analysis and Discussion

Table 6.3 shows some statistical information collected during the testing stage. Table 6.3 shows that the total cases of Idgham are 1,818 and the total cases of Iqlaab are 200. The Idgham of Nuun Saakina and Tanween is the highest to occur among all Idgham forms. This clearly shows that Idgham occurred more than Iqlaab in MSA. Among the identical letters, our results show that Idgham of two consequent Lam (ل) has the highest frequency to occur in Arabic language. Table 6.3 shows that Lam (ل) followed by Lam (ل) occurred 49 times, based on our corpus.

Even though 2,018 compound words have been found in the corpus, only 1,639 compound words have been actually added to the dictionary after excluding the repetition.

Tables 6.4–6.6 provide samples of the recognition results of the baseline and the enhanced systems. The samples show how the added compound words help to improve the performance.

During recognition, 117 compound words were provided by the enhanced dictionary. After recognition process, these compound words were switched back to its separated form. However, this does not mean that they were misrecognized in the baseline system. Many of them were correctly recognized in the baseline system as separated words.

For more clarification, we carefully analyzed the recognizer outputs. We measured the percentage of recognition in both systems among all tested files. Table 6.7 shows that the proposed method leads to improvement in some speech files and, however, to decrease in performance in others.

Figure 6.1 demonstrates the information provided in Table 6.7 in Pie chart.

We mentioned in Table 6.7 that some correctly recognized words in the baseline were misrecognized in the enhanced system. The following are two illustrative examples listed as the following order: original speech to be tested, baseline system recognition results, and enhanced system recognition results, respectively.

- فَسَيُتَرَكُ قَرَارُ التَّخصِيص لِهَيئَةِ سُوقِ المَال
  fasayutraku qararu altakhsiisi lihy'ti swq 'lmal
- فَسَيُتَرَكُ قَرَارُ التَّخصِيص لِهَيئَةِ سُوقِ المَال
  fasayutraku qararu altakhsiisi lihy'ti swq 'lmal
- فَسَيُتَرَكُ قَرَارُ التَخصِيص فِي لِهَيئَةِ سُوقِ المَال
  fasayutraku qararu altakhsiisi lihy'ti swq 'lmal

- الّتِي لَدَيهَا صَرّافَاتُ آلِيَّة أو مُصدِرَة لِلبِطَاقَاتِ الذَّكِيَّة
  'laty ladyha Sarafatun 'liya 'w muSadira lilbitaqati aldhakiya
- الّتِي لَدَيهَا صَرّافَاتُ آلِيَّة أو مُصدِرَة لِلبِطَاقَاتِ الذَّكِيَّة
  'laty ladyha Sarafatun 'liya 'w muSadira lilbitaqati aldhakiya
- الّتِي لَدَيهَا صَرّافَاتُ آلِيَّة أو الذِين مُصدِرَة لِلبِطَاقَاتِ الذَّكِيَّة
  'laty ladyha Sarafatun 'liya 'w **'ldayn** muSadira lilbitaqati aldhakiya

We noticed that most of the errors that occur in the enhanced system (i.e., they are correct in the baseline) have no relation with compound words. None of them made cross-word transformation process. We believe that the source of these errors is the language model as it is recalculated according to the enhanced corpus

Table 6.3 Rules usage in the entire transcription corpus

Rule	Final letter Of first word **(unvowelled)**	Initial letter Of second word **(vowelled)**	Usage times
1	A letter	Identical with the previous letter	
	baa' / ب	baa' / ب	17
	taa' /ت	taa' /ت	38
	thaa' /ث	thaa' /ث	0
	jiim / ج	jiim / ج	0
	Haa' / ح	Haa' / ح	0
	khaa' / خ	khaa' / خ	0
	daal / د	daal / د	2
	dhaal / ذ	dhaal / ذ	0
	raa' / ر	raa' / ر	16
	zaay /ز	zaay /ز	1
	siin / س	siin / س	7
	shiin /ش	shiin /ش	0
	Saad /ص	Saad /ص	0
	Daad /ض	Daad /ض	0
	Taa' /ط	Taa' /ط	0
	Zaa '/ظ	Zaa '/ظ	0
	'ayn /ع	'ayn /ع	18
	ghayn / غ	ghayn / غ	0
	faa' /ف	faa' /ف	12
	qaaf / ق	qaaf / ق	3
	kaaf / ك	kaaf / ك	0
	laam /ل	laam /ل	49
	miim / م	miim / م	42
	nuun / ن	nuun / ن	0
	haa' / ه	haa' / ه	0
			=====
			205
2	Nuun Saakinah and Tanween	yaa' /ي raa' /ر miim /م laam /ل waaw /و nuun /ن	1531
3	Nuun Saakinah and Tanween	baa' / ب	200
	A letter	A close in pronunciation letter	
4	taa' /ت	daal / د	25
5	taa' /ت	Taa' / ط	4
6	daal / د	taa' / ت	32
7	baa' / ب	miim / م	14
8	dhaal / ذ	zaay / ظ	0
9	kaaf /ق	kaaf /ك	1
10	laam /ل	raa' / ر	6
Total			2018

Table 6.4 Idgham case: unvoweled nuun (nuun Saakinah) followed by raa'

Original speech to be tested	 بَعدَ شَهرٍ وَاحِدٍ مِنْ رَفعِهَا لِلحَظر b'd shahrin wahidin min raf'iha lilhazr
As recognized by the baseline system	بَعدَ شَهرٍ وَاحِدٍ رَفعِهَا لِلحَظر b'd shahrin wahidin raf'iha lilhazr
As recognized by the enhanced system	بَعدَ شَهرٍ وَاحِدٍ مِرَّفعِهَا لِلحَظر b'd shahrin wahidin mirraf'iha lilhazr
Final output after decomposing the merging	بَعدَ شَهرٍ وَاحِدٍ مِنْ رَفعِهَا لِلحَظر b'd shahrin wahidin min raf'iha lilhazr

Table 6.5 Idgham case: unvoweled nuun (nuun Saakinah) followed by miim

Original speech to be tested	 تُجبِرُهَا عَلَى الاِبتَعَادِ عَنْ مَلاعِبِ tujbiruha 'ala al'bti'adi 'an mula'ib
As recognized by the baseline system	تُجبِرُهَا الاِبتِعَاد عَنْ اللاَّعِبِ tujbiruha al'bti'adi 'an 'lla'ib
As recognized by the enhanced system	تُجبِرُهَا عَن الاِبتِعَاد عَمَّلاعِبِ tujbiruha 'an al'bti'adi 'ammula'ibi
Final output after decomposing the merging	تُجبِرُهَا عَن الاِبتِعَاد عَنْ مَلاعِبِ tujbiruha 'an al'bti'adi 'an mula'ibi

transcription. Recalculation of the language model probabilities according to the new transcription presented a major change in the *n*-gram probabilities. Table 6.8 shows the total count of 1-grams, 2-grams, and 3-grams of the language model for both the baseline system and the enhanced system. So, the new language model might be biased to some word sequences on the account of others.

According to the data provided in Table 6.8, we found that *n*-grams have been increased according to the compound words. This increase in the total of *n*-grams will provide an opportunity for enhancement. Saon and Padmanabhan (2001) showed mathematically that compound words will enhance the performance.

Table 6.6 Iqlaab case: unvoweled nuun (nuun Saakinah) followed by baa'

Original speech to be tested	لِلاشْتِرَاكِ فِي المَزَادِ العَالَمِيِّ مِنْ بَيْنِ lil'shtiraki fy 'lmazadi 'l'alamyi min bayn
As recognized by the baseline system	لِلاشْتِرَاكِ فِي المَزَادِ العَالَمِيِّ بَيْنِ lil'shtiraki fy 'lmazadi 'l'alamyi bayn
As recognized by the enhanced system	لِلاشْتِرَاكِ فِي المَزَادِ العَالَمِيِّ مِمْبَيْنِ lil'shtiraki fy 'lmazadi 'l'alamyi mimbayni
Final output after decomposing the merging	لِلاشْتِرَاكِ فِي المَزَادِ العَالَمِيِّ مِنْ بَيْنِ lil'shtiraki fy 'lmazadi 'l'alamy min bayni

Table 6.7 A comparison between the baseline and the enhanced system

Among the 1,144 speech files	Recognized files in (baseline, enhanced)
1,047 Speech files (91.5%)	Both systems (the baseline and the enhanced) agreed upon recognition of these files, either correctly or incorrectly (we ignored light diacritic differences)
23 Speech files (2.01%)	Recognized correctly in the baseline system but are not in the enhanced system
74 Speech files (6.46%)	Recognized correctly in the enhanced system but are not in the baseline system

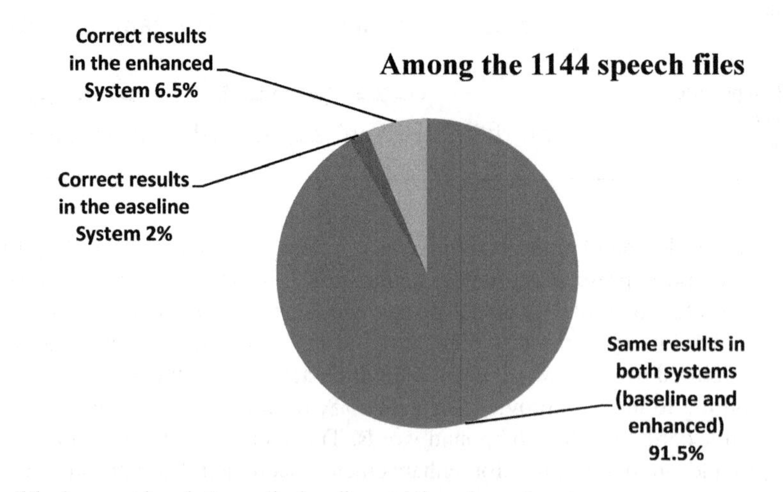

Fig. 6.1 A comparison between the baseline and the enhanced system

Table 6.8 *n*-Grams of both systems (baseline and enhanced)

System	1-grams	2-grams	3-grams
Baseline	14,234	32,813	37,771
Enhanced	15,873	37,852	45,858

Table 6.9 Samples of indirect improvements by the language model

Original speech to be tested	مِن الغَازِ الإيرَانِيِّ إلَى الهِند min 'lghaz 'l 'iirany 'la 'lhind وَمُمَثِّلِينَ عَن عَدَدٍ مِن الدُّوَلِ الأورُوبِيَّة wamumathiliina 'an 'adadin min 'lduwal 'l'wrwbiya بِمَرَضِ جُنُونِ البَقَر bimaraD junwn 'lbaqar
As recognized by the baseline system	مِن الغَازِ الإيرَانِيِّ إلَى الحَلبَة min 'lghaz 'l 'iirany 'la 'lHalaba وَمُمَثِّلِين عَن إنَّ الدُّوَلِ الأورُوبِيَّة wamumathiliina 'an **'na** 'lduwal 'l'wrwbiya فِي بِمَرَضِ جُنُونِ التَّقَاعُدِ fy bimaraD junwn 'ltaqa**'ud**
As recognized by the enhanced system	مِن الغَازِ الإيرَانِيِّ إلَى الهِند mn alghaz alayrany ala alhnd وَمُمَثِّلِينَ عَن عَدَدٍ مِن الدُّوَلِ الأورُوبِيَّة wamumathiliina 'an 'adadin min 'lduwal 'l'wrwbiya فِي بِمَرَضِ جُنُونِ البَقَر fy bimaraD junwn 'lbaqar

They demonstrated that the compound word has the effect of incorporating a trigram in dependency in a bigram language model, as an example. Generally, compound words are most likely to be correctly recognized more than separated words. Consequently, correct recognition of a word might lead to another correct word through the enhanced *n*-gram language model. In contrast, misrecognition of a word may lead to another error in the word sequence and so on.

Table 6.9 gives an example of the robustness we described above which leads to indirect enhancement. It shows the enhancement of a sentence that has no transformation process, i.e., the enhancement is there while there is no cross-word phenomenon in the sentence to be tested.

We can conclude that the new language model, generated by the expanded transcription, introduces both improvement and ambiguity. This is why 2.01% among testing files were misrecognized in the enhanced system.

Although our method enhanced the overall performance of the speech recognizer, however, we have observed a few cases in which the application of the method created misrecognition cases, which were properly recognized before. The performance enhancement together with the introduction of new errors are related to the language model's n-grams recalculation. It is clear that the more cross word cases we append to the language model, the more cross-word errors we remove from the error set, though not in a linear proportion. In the mean time, the modification in the language model may negatively change the n-gram probabilities of some words, leading to new recognition errors. This phenomenon may raise a question for further research about possible optimality of the modified language model, a language model which makes the best compromise between removing the cross-word errors, and generation of other errors.

The great impact on the perplexity could be understood in two ways: first, the robustness that occurred in the language model increases the probability of the testing set $W = w1, w2,\ldots,wN$, therefore reducing the perplexity according to the perplexity formula (explained in Sect. 6.2):

$$\mathrm{PP}(W) = \sqrt[N]{\frac{1}{\mathrm{P}(w1, w2, \ldots, wN)}}.$$

According to the formula, it is clear that increasing P will reduce the PP. Second, the 1,639 compound words added to the transcription as new words have an extremely low perplexity. For example, consider the two words (من) and (بعد). These two words have an average certain perplexity. When the compound word (ممبعد) is represented in the language model, it will share others with its low perplexity, so reducing the overall perplexities.

Finally, our method was implemented as a preprocess step to extend the span of the dictionary and the language model. The training stage has not evolved, i.e., the acoustic model of all training utterances has not been changed during the experiment.

References

Gallwitz F, Noth E, et al (1996) A category based approach for recognition of out-of-vocabulary words. In: Proceedings of fourth international conference on spoken language, 1996. ICSLP 96

Jelinek F (1999) Statistical methods for speech recognition, Language, speech and communication series. MIT, Cambridge, MA

Jurafsky D, Martin J (2009) Speech and language processing, 2nd edn. Pearson, NJ

Plötz T (2005) Advanced stochastic protein sequence analysis, Ph.D. thesis, Bielefeld University

Saon G, Padmanabhan M (2001) Data-driven approach to designing compound words for continuous speech recognition. IEEE Trans Speech Audio Process 9(4):327–332

Closing Remarks

The proposed knowledge-based approach achieved feasible improvement for cross-word variation modeling. Mainly, two MSA phonological rules were applied: the Idgham and Iqlaab. The experimental results clearly showed that the Idgham occurred more than Iqlaab in MSA. So, the Idgham rules dominate the generation of the cross-word variants. The significant enhancement we achieved has not only come from the cross-word pronunciation modeling in the dictionary, but also indirectly from the recalculated n-gram probabilities in the language model. One of our major conclusions is that decoding process in ASRs using Viterbi algorithm gives better results with long words (i.e., compound words), and this observation still needs to be utilized to improve the performance in the future work.

Future work may also consider additional phonological rules such as the Hamzat Al-Wasl and having two unvoweled consecutive letters. Further work is also needed to study the contribution of each phonological rule in order to select efficiently the most frequent ones or the most effective ones. Finally, one more thing to be studied is the effect of retraining the acoustic model based on a modified training corpus with the proposed models of cross-word pronunciation variations.

D. AbuZeina and M. Elshafei, *Cross-Word Modeling for Arabic Speech Recognition*,
SpringerBriefs in Electrical and Computer Engineering,
DOI 10.1007/978-1-4614-1213-7,

Appendix
Arabic Terminologies

Al-Alta'rif The determiner (ال).

Damma An Arabic short vowel (ُ), pronounced like (u).

Dammatan Two Damma (or doubling of Damma), pronounced like (n). Also called Tanween of Damma.

Fatha An Arabic short vowel (َ), pronounced like (a).

Fathatan Two Fatha (doubling of Fatha), pronounced like (n). Also called Tanween of Fatha.

Hamzat Al-Wasl It is an extra Hamza that helps to start pronouncing an unvoweled letter in Arabic continuous speech.

Idgham Also called geminating or assimilation, it is a merging of two consecutive letters of the second type letter.

Idgham almutajanisan It is a merging between two consecutive different letters that are close in pronunciation. Some of these cases include: taa'/تْ and daal/د, taa'/تْ and Taa'/ط, dhaal/ذْ and Zaa/ظ, qaaf/قْ and kaaf/ك, and laam/ل and raa'/ر.

Idgham almutmathlan It is a merging between two consecutive identical letters shown in the following list {ب, ت, ث, ج, ح, خ, د, ذ, ر, ز, س, ش, ص, ض, ط, ظ, ع, غ, ف, ق, ك, ل, ن}. The rule means that any unvoweled Arabic letter followed by the same Arabic voweled letter will be doubled in a single merged word. Note that {ا, و, ي} are not included in the list.

Iqlaab It is a replacement of unvoweled nuun (Nuun Saakinah <>نْ) or Tanween (ٌ، ً، ٍ) that comes before voweled baa' (ب) by unvoweled miim (Miim Saakinah <>مْ).

Kasra An Arabic short vowel (ِ), pronounced like (i).

Kasratan Two Damma (doubling of Kasra), pronounced like (n). Also called Tanween of Kasra.

Nuun Saakina An unvoweled nuun symbolized as (نْ).

Shadda It is a doubling of consonant and is symbolized as (ّ).

Shamsi group Arabic letters include taa', thaa', daal, dhaal, raa', zaay, siin, shiin, Saad, Daad, Taa', Zaa', laam, and nuun.

Sukun Absence of vowel, symbolized by (ْ).

Ta'al marbouta It is an Arabic letter symbolized as (ة) and shown at the end of the words.

Tanween Includes any one of Dammatan, Fathatan, or Kasratan. It is symbolized as (ـٌ ، ـً ، ـٍ).

Index

9781461412120